WILDLIFE CONSERVATION IN AFRICA

WILDLIFE CONSERVATION IN AFRICA

A Scientific Approach

S.S. AJAYI

ACADEMIC PRESS
An imprint of Elsevier

Academic Press is an imprint of Elsevier
125 London Wall, London EC2Y 5AS, United Kingdom
525 B Street, Suite 1650, San Diego, CA 92101, United States
50 Hampshire Street, 5th Floor, Cambridge, MA 02139, United States
The Boulevard, Langford Lane, Kidlington, Oxford OX5 1GB, United Kingdom

Library of Congress Cataloging-in-Publication Data
A catalog record for this book is available from the Library of Congress

British Library Cataloguing-in-Publication Data
A catalogue record for this book is available from the British Library

ISBN 978-0-12-816962-9

For information on all Academic Press publications
visit our website at https://www.elsevier.com/books-and-journals

Publisher: Charlotte Cockle
Acquisition Editor: Anna Valutkevich
Editorial Project Manager: Laura Okidi
Production Project Manager:
Punithavathy Govindaradjane
Cover Designer: Matthew Limbert

Typeset by SPi Global, India

Contents

About the Author

Professor **Emeritus S.S. Ajayi** was born in Mopa, Mopamuro Local Government Area, Kogi State of Nigeria, in May 1944. He attended Sudan Interior Mission (SIM) Central School, Mopa (1951–57) and SIM (ECWA) Titcombe College, Egbe, Kogi State (1958–62). He won a Northern Nigeria Government scholarship award for higher school certificate course at Gindiri Boys' High School, Plateau State (1963–64). He also won the second Northern Nigerian Government scholarship award to Ahmadu Bello University (ABU), Zaria, where he obtained a BSc honors in zoology (1965–68). Professor S.S.Ajayi also won the third scholarship award—United Nations/Food and Agriculture Organization (FAO/UNDP) Fellowship to attend University of Edinburgh, in the United Kingdom, for an MSc program by research in wildlife and range management and also the College of African Wildlife in Tanzania, East Africa (1969–71).

His MSc FAO fellowship program at University of Edinburgh Scotland and his practical orientation training course at the college of African wildlife in Tanzania were rated by FAO fellowship division as one of the most successful it has undertaken. He won the British Commonwealth Technical Cooperation Scholarships twice to attend Central Institute of Fisheries Education in Bombay (Mumbai), India (1986 and 1989). He won the British Council Fellowship in 1985 for a study tour of some British institutions engaged in research and development in wildlife and fisheries management in the United Kingdom. Through a master-plan proposal for the development of wildlife and fisheries management at the University of Ibadan, Professor Ajayi won six British Council Scholarships on which the core foundation staff of the fist Department of Wildlife and Fisheries Management in Africa were trained.

Professor Emeritus Ajayi obtained his PhD degree at the University of Ibadan in September 1974. His PhD principal supervisor was Professor Laurence Roche, an Irish-Canadian professor of forestry, who was the head of Department of Forestry at the University of Ibadan (1973–76). His cosupervisor was Professor J.K. Loosli, a renowned emeritus professor at the Department of Animal Science, University of California, the United States. Professor Loosli was a Rockefeller Foundation contract staff in the department of Animal Science, the University of Ibadan (1972–74). The external examiner was Professor L.B. Halstead (DSc) of the Department of Zoology, University of Reading, the United Kingdom.

Professor Emeritus Ajayi was appointed assistant training fellow at the University of Ibadan on December 8, 1968. He was promoted to lecturer II in 1972, lecturer grade I in 1974, senior lecturer in 1975, and reader in 1978. He was promoted the first African professor of wildlife management on October 1, 1980. He was the pioneer head of Department of Wildlife and Fisheries Management (1981–88) and an exceptional dean, College of Environmental Resources Management, University of Agriculture Abeokuta (UNAAB)—1988–90. Professor Ajayi was one of the first four professors that established the Federal University of Agriculture, Abeokuta (1988–90).

Professor Emeritus S.S. Ajayi developed a wildlife program from half of a three-unit course in 1972 to become full-degree programs of wildlife management and fisheries management in 1981 when the Department of Wildlife and Fisheries Management was created, and as far as it is known, it was the first of its kind in Africa.

With the collaboration of his colleagues, the two academic programs that have received full National Universities Commission (NUC) accreditation represent 50% of four degree programs in the Faculty of Agriculture and Forestry at the University of Ibadan up to 2009 when he retired mandatorily from the service of the University of Ibadan.

Professor Emeritus S.S. Ajayi was a recipient of the United Nations Environment Programme (UNEP) "Global Roll of Honour" award for outstanding practical achievements in environment protection. Professor Ajayi was honored by UNEP in London on June 4, 1994 at Queen Elizabeth II Conference Centre specifically for the following achievements:

For the creation of the first functional, autonomous Department of Wildlife and Fisheries Management and subsequently became the pioneer head of the new department.

For developing scientific approach to wildlife conservation through Africa.

At the national level, Professor S.S. Ajayi was the project leader of an expert group that produced an action plan for the conservation of all renewable natural resources in Nigeria.

He developed the technologies for the domestication of three species of African wildlife: (a) the African giant rat, (b) the grass cutter, and (c) the Helmeted guinea fowl.

Professor Emeritus S.S. Ajayi was the vice president of the World Conservation Union—West African Sustainable Specialist Group (1999–2003).

Professor Emeritus S.S. Ajayi worked as a consultant to United Nations/ Food and Agriculture Organization (UN/FAO) in the following countries: Liberia, Ghana, Ivory Coast, Cameroon, Zambia, Zimbabwe, Botswana, Malawi, and Mozambique from 1979 to 2002. In 1992 UN/FAO wrote the following report on Professor Ajayi:

> *Professor Emeritus Ajayi of University of Ibadan has done some excellent work for us (FAO) on Consultancy basis. He remains one of the First Consultants we call upon we need one.*

On August 5, 2010 he was honored by the Fauna Conservation Society in Nigeria as "the father of wildlife and fisheries in Africa."

On December 5, 2013 Professor Ajayi won the highest award that the Nigerian government can present to its citizens—the Prestigious Nigerian National Order of Merit (NNOM) in Science. This award was the highest national prize for academic and intellectual attainment. He was also presented with a cash award of N10 million in recognition of his significant intellectual works in wildlife management and environmental protection. The last portion of Professor S.S. Ajayi's citation thus reads "It is this humble, but highly talented international figure who has held the banners of our dear country Nigeria high that was given this National Award."

PART 1

Introduction

CHAPTER 1
Introduction

What is conservation?

The science of wildlife conservation is of a relatively recent origin. As a pioneer, the objective of this book is to unfold the potential and significance of wildlife in the economies of African countries. It also presents the problems of wildlife conservation, principles, and scientific approach to wildlife conservation in sub-Saharan Africa. Since this book dwells on "conservation" as a scientific tool for maintenance and utilization of wildlife in perspective, it may be instructive to familiarize readers with this key word for two main reasons. First, there has been some confusion over the use of the word, and secondly, there is the lack of realization that in fact all forms of national development itself are based on conservation.

In the search for the true meaning of "conservation," there are three definitions that I find acceptable in terms of theory and application to renewable natural resources and that foster the purpose of integrated multiple land use. The first definition is culled from Macmillan Encyclopedia Dictionary as follows:

> *The rational use of the earth's resources so that life can be sustained indefinitely.*

The second definition is from *Tropical Forest Ecosystems* by UNESCO (1978), and it states as follows:

> *The utilization of resources, that is, a sustainable utilization, resting on ecological principles, and therefore including agricultural, forestry or pastoral development.*

The last definition is that of Gabrielson (1970):

> *In its broadest social aspects, conservation of organic resources means restoring to the highest possible level and maintaining at a state of high productivity, those resources, including wildlife, that can be used on a crop basis to sustain human society.*

In the above definitions, there is an explicit recognition that man disturbs the fundamental harmony and balance of nature by his activities. Throughout the history of land development all over the world, man at some critical moments resulting from disruption evolves an effort to

Wildlife Conservation in Africa
https://doi.org/10.1016/B978-0-12-816962-9.00001-6

ensure his security. "Conservation" would therefore appear to be a security-ensuring activity that often minimizes uncertainties in the provision of our basic necessities of life resulting from man's destructive activities. These definitions are, therefore, based on the ecological philosophy of conservation.

One of the basic concepts of ecological principles of conservation, whether we are fishermen or wildlifers or miners, is the fact that no natural resource is inexhaustible. The wildlife of the forests, the fish of our rivers, and the minerals in the ground if exploited indiscriminately would be exhausted. There would be a time when they will be no more! The recognition of this basic fact leads to the concept of "preservation" as the first step in "conservation." If the resource base is not preserved, there would be nothing to utilize or manage.

Conservation movements in America started as a result of the land use mismanagement that characterized the early settlement of the continent. This phenomenon naturally led to a threat of scarcity of available natural resources, the existence of the United States as a nation, as well as the destruction of various facets of nature's beauty. Furthermore, it led to the emergence of two basic conservation philosophies early in the 20th century. One group of conservationists promoted "preservation" or "nonuse," while the other promoted natural resource development or "rational use" (Gabrielson, 1970).

In the history of the United States, the first conscious effort to effect a change in the area of natural resource management was by a German-trained forester named Gifford Pinchot, the man generally recognized as the father of conservation movement in America. Pinchot's principle of conservation connoted "use through systematic management." He was opposed to established forest reserves that were "preserved" and not "used." He believed that it is only when a forest is preserved and used under a systematic form of management based on sound ecological principles that the forest is conserved. Indeed, the term "conservancies," as it applied to government forests in India in those days, appealed to Pinchot so much that he used "conservation" to describe his public campaigns for forestry and other natural resource sector in 1907. Thus, the evolution of conservation that started basically as mere preservation of the environment crept gradually to include "rational use" by man.

This brief historical perspective has been cited to show two things: first is that the two principles of preservation and conservation were born to protect the American nation's declining resources. Secondly, the concept and principles that led to the philosophy of conservation were agrarian and purely ecological.

From this point, the metamorphosis of conservation principles took a new turn. It accepted development as part of man's culture and made it to represent the second environment or "frontier" in conservation. For the purpose of this book, it is assumed that the agrarian ecological factors such as land, forest, wildlife, and fish belong to the horizontal plane of "frontier" of conservation. On the vertical plane, we have the new environment of developmental factors such as economic development, science, technology, and research as they affect the land and the welfare of man. The problems that are associated with the feedback of various forms of cultural development, be it economic, research, or technology, are so real that these later planes affect the welfare of man, thus become an integral part of the principles of conservation.

One fact that is partly responsible for America's success in various forms of development was that they recognized the principles of "what goes up must come down," that is, for all forms of development, land and its resources are the ultimate recipients and therefore must be tuned to "carry the load" eventually. They have endeavored to ensure that land resources would carry development without disharmony among these resources themselves or between the resources and man.

Take industrial development as an example. Most industries are associated with industrial wastes. It is not enough to campaign for industrial development and create a favorable economic atmosphere for investors, without thinking about the wastes associated with such developments. How are these wastes associated with such developments? How are these wastes tolerated by man, animals, and plants? What damages do they cause, and how can these be minimized or eliminated? The need for a reconciliation of the concepts of development was tightly noted by Fisher (1968):

> *We shall have to explain the natural environment and its several parts more comprehensively and systematically, looking further ahead, and paying more attention to the interrelations between that environment on one hand and the cultural environment of research, technology, economic development and human welfare on the other hand. The problems call for solutions now.*

This concept is in consonance with the fact that development carries with it two environments of conservation—the natural and the cultural—and the problems associated with them must be reconciled. In the developing countries of Africa, we cannot afford to ignore these two frontiers of conservation.

It must be stated that any developing country has more to learn about conservation principles and practice, specific details of legislations, campaigns, government, and private involvement in conservation from the United States more than any other country worldwide.

Furthermore, it must be stressed that one of the basic principles of conservation is preservation and that preservation is not synonymous with conservation, but the first step toward conservation. It is also assumed that the concept of two frontiers of conservation is clear, that is, the natural (ecological) and the cultural environment (science, technology, and economic development).

It is perhaps pertinent to point out that there are numerous references in the Bible such that make conservation an integral phenomenon to our physical and spiritual lives. The most striking reference to conservation of renewable natural resources is found in Genesis 7:2–3. Here, the very principle of conservation was laid down by God Himself, and Noah was the first to practice it. It reads thus:

> *Of every clean beast, thou shall take to thee by sevens, the male and his female; and of the beasts that are not clean by two, the male and his female, to keep seed alive upon all the face of the earth.*

And Noah did according unto all that the Lord commanded him (Genesis 7:5).

In Genesis 7:13, Noah and his children did not leave their wives behind—because to make themselves renewable, they needed females.

This was the very conservation principle and practice that preceded the great flood. The Almighty God even in anger was cautious enough to preserve Noah and his family first because they were upright before Him and second because He wanted them to perpetuate their kinds. If God had preserved human beings alone through Noah's family, with disregard to beasts and fowls, the world after the flood would have been without lower animals! We would have had today no wildlife and no domestic animals, and therefore, there would have been no hunters, no zoologists, and no animal breeders or veterinarians. It also follows that there would be no meat or fish for us to eat or milk to drink.

Why is this book about conservation of African wildlife?

The African continent is endowed with the highest diversity of wildlife in the world. However, it is because of the alarming rate at which African wildlife is being lost daily to unregulated exploitation, and illegal trade in endangered species is a threat to the future of wildlife in Africa. For example, the declining trend in wildlife in West Africa, which was about 214 kg/

km^2 for herbivores in several national parks in 1978, is now less than 5% of the potential carrying capacity of their habitats.

The highly priced species such as gorilla, chimpanzee, hippopotamus, leopard, lion, and giraffe face the threat of extinction in West and Central Africa. Indeed, the last rhinoceros was reportedly killed in Nigeria in what is now known as Bauchi State in 1936. The total number of rhinoceros in Africa is now estimated at less than 3000 and lions at 20,000. Furthermore, in the vast unprotected savanna regions of Africa, wildlife density is estimated at less than 1% of their carrying capacity.

The negative trend of wildlife population all over sub-Saharan Africa is a wake-up call for African governments to make wildlife conservation a priority in governance in order to rescue this vital economic, cultural, and medicinal asset from extinction.

The alarming rate at which African wildlife is being decimated is a wake-up call for all-embracing, comprehensive approach to conservation of this vital resource by African countries. That's why this book dwells on the potential and actual value of wildlife in Africa's economies; their cultural and medicinal values; its declining status; and the principles, strategies, and programs of conservation of African wildlife.

Scope of this book

The survey of the importance of forest wildlife in West Africa was carried out by the author in six West African countries: Cameroon, Nigeria, Benin, Ghana, Ivory Coast, and Liberia in 1979. These are countries that possess extensive tropical forests and where wildlife is recognized as assets, in terms of food and animal products. In order to have a balanced view of the various subject areas, it was recognized that wildlife utilization and management practices in West Africa are to some extent influenced by colonial background. On this basis, three Francophone countries (Cameroon, Benin, and Ivory Coast) and three Anglophone countries (Nigeria, Ghana, and Liberia) were visited, and on the spot assessments of the various aspects of wildlife utilization were made. The study was confined to tropical forest areas only and limited to the utilization of terrestrial mammals. A generalized questionnaire was used as a guide for collating information from the six countries visited.

While this book is meant primarily as a textbook for students of wildlife conservation in tertiary institutions, it will also serve as a guide for wildlife extension workers in tropical Africa and rural dwellers who are most affected by the status of this resource. It is meant also to assist administrators and planners to focus their attention on wildlife conservation because of its values in rural and national economies. In competing for government funds, development planners in West Africa, as in many other parts of the world, frequently rely on the "cost/benefit" ratios of projects for allocation of funds. Traditionally, wildlife management and nature protection in general have been part of Government Forestry Departments in most African countries. However, forestry and wildlife managers have suffered from dearth of information on this subject due to neglect and also to the fact that the gestation periods of conservation projects are often too long for development planners to appreciate and prioritize. It is hoped that the information in this book will serve the purpose of drawing the attention of administrators to the socioeconomic and conservation benefits of wildlife and other natural resources.

PART 2

The importance of wildlife in the economies of African states

CHAPTER 2

Utilization of forest wildlife in West Africa

Introduction

Wildlife is generally acknowledged in Africa as a valuable source of protein and revenue. However, there are few hard facts in the form of comprehensive published information on its pattern of utilization and its nutritive and economic importance. Moreover, forest wildlife in Africa, which has from time immemorial provided man with a significant portion of animal protein intake, has received very little attention from scientists, administrators, development planners, and financiers of development programs. In West Africa, where the tropical forest occupies a large proportion of the total vegetation cover, the role of wildlife as a source of food is probably more important than elsewhere in Africa. This is because of the prevalence of tsetse flies (*Glossina* species) that make large-scale production of livestock difficult. As far as it is known, there has not been a comprehensive collation of quantitative information on the forest wildlife of West Africa as a source of food and by-products. Furthermore, evaluation of traditional methods of production, harvesting, and marketing and the contribution of forest wildlife to local communities in socioeconomic terms have not received much attention.

Chapters 2–6 of this book are collections of information and results of research projects by the author on the utilization of forest wildlife in West Africa for food and animal products. It is designed to evaluate traditional methods of utilization of wildlife, with particular reference to mammals. It also includes an assessment of traditional processing methods for bushmeat and by-products in rural economies and the nutritive value of forest species that are significant as food items. Also reported here are examinations of the wildlife management practices in several West African countries, with a view to exploring possibilities for integrated use of wildlife and food crops in the broad concept of traditional and modern farming with wildlife and forestry in line with the practices of agroforestry.

Wildlife Conservation in Africa
https://doi.org/10.1016/B978-0-12-816962-9.00002-8

While recognizing the role of wildlife as a source of meat, there are ways of managing these resources together with food crops so that the former can be exploited on a rational basis, in order that their existence and food habits on farms and in forests would not cause "economic damage." These chapters attempt to elucidate this concept with a collation of various on-going research projects in West Africa and to identify gaps in knowledge on the status, biology, and methods of management. This approach will facilitate the drawing up of research priorities directed at more efficient utilization of forest wildlife.

Important forest mammals utilized for food and animal products

Species utilized for food and animal products

Asibey (1974a,b) stated that protein deficiency and malnutrition were prevalent in Africa, south of the Sahara, partly because wildlife that used to be man's main source of animal protein is now in short supply. This is partly due to human population pressure. Overhunting has depleted wildlife in most parts of West Africa, and therefore Africans in this subregion cannot afford to be as selective as they were about half a century ago. People nowadays eat whatever meat is available. In southern parts of Nigeria and Ghana, maggots, various insects, and agama lizards are eaten (Asibey, 1978). In the tropical forest region, where, because of environmental hazards, domestic livestock cannot thrive as in the savanna regions, wildlife is and will continue to be of paramount importance because with the naturally low-protein diets of the inhabitants, every gram of animal protein is important. It is therefore pertinent to state that practically, all forest mammals are eaten in West Africa—ranging from bats, squirrels, galagos, and monkeys to antelopes and elephant.

Uses of animal products

The term "animal products" cover any part of the animal body, including its droppings that are used by people other than for food. The following are several ways in which animal products are utilized:

Ivory of elephants, teeth of the hippopotamus, and skins of big cats are used for ornamental purposes.

Skins of antelopes such as yellow-backed duiker, bushbuck, bongo, and colobus monkeys are used for decorations and other valuable household materials, such as ladies handbags and purses.

In West Africa, especially in Southern Nigeria and Southern Cameroon, any part of wild animal may be used for traditional medicine (Ajayi, 1978). In some cases, a whole rodent may be used as an ingredient of some medicinal preparations or for witchcraft. In other cases, only the parts of animals including feces, skins, bones, and meat are used for a variety of medical concoctions.

Forest mammals are also widely used in scientific research. Primates are used in biochemical, physiological, and medical research because of their close resemblance to man. In fact, the "rhesus factor" in blood types and blood clotting was first discovered in monkeys, and the findings were utilized for man. Primates and other mammals are also used to test new drugs, because their physiological processes are close to humans.

Pattern of bushmeat production, preservation and marketing

Hunting tools

Hunting tools evolved in West Africa from simple traps (spears, bows, and arrows) to guns, rifles, and even automatic weapons. In most West African countries, forest mammals are still hunted with snares, traps, spears, bows and arrows, poisoned baits, and nets. Dogs are also used in community traditional group hunting. The introduction of dane guns, shotguns, and rifles as a result of civilization also gave rise to a corresponding higher demand for wild animal protein. However, it is interesting to note that both the primitive and modern hunting tools are still employed in West Africa today. The following are among hunting techniques identified in this part of Africa.

Hunting techniques

1. *Trapping and snaring*

Trapping and snaring is a hunting technique that is usually associated with subsistence farming and is mainly used for mammals, small, medium, and big game. They consist of trap-fence line (for small game), gin traps (for small game), and snares and pitfalls (for both small- and medium-size game). Snares and traps may be baited, depending on the type of animals desired to be killed. They are usually set for animals that have regular routes especially where well-defined tracks exist. This hunting technique is common in all over West Africa.

2. *Individual professional hunting with dane guns and shot guns*

This hunting technique involves a professional hunter going out with a dane gun to hunt, frequently with an assistant, to carry his kill. The kill is usually eaten by the hunter's family, and when a big animal is killed, it is either sold fresh to local consumers or to a local collector who smoke-dries it for sale to a retailer.

Night hunting is also practiced in a similar way. The hunter who bears the arm carries a headlamp that dazzles the animal, and it is the reflection from the animal's eyes that betrays it. Using this method, a hunter may kill animals like the size of bushbuck or a bigger game such as buffalo. In most parts of West Africa, any hunter who kills a buffalo is usually elevated among the hierarchy of hunters in his rural community, because of the risks involved in using a semiprimitive weapon for a rather big and aggressive animal.

3. *Community hunting for commercial purposes*

Traditional hunting tribes may have little to do after the farm harvests at the early part of the dry season. They set out in groups ranging from 3 to 10 and may be up to 20 in number including their youngsters. They may travel long distances and invade anywhere that game is relatively abundant, including the rain forests. In Nigeria, the Keberi, Zuru, and Dankari Kari tribes of the North Western Zone are noted for this type of hunting drive, while in Liberia the Sapo and Kran tribes also use this technique extensively. Such a hunting spree may last several weeks, during which their kills are smoke-dried and sold to retailers who in turn sell to the markets. This type of hunting is wasteful as animals are indiscriminately killed for cash value.

4. *Annual festival community hunting*

Another type of community hunting in West Africa is of a festive nature. It is often an annual event where villagers go out during the dry season (dry season is regarded as a closed season) to hunt on a particular area of land, which is the traditional property of that particular village. A village is forbidden to hunt on the property of another village. When hunters reach the hunting ground, they space themselves out in a few yards from each other, and dogs are assigned to comb the bush and drive the animals out. The participants make a lot of noise as they cut their way through the bush. When they see an animal, there is a chase, and the animal is run down and killed. The kill is usually shared among participants, but some are reserved for the village head as his traditional right. The method of hunting with defined close season whereby local people respect the tradition not to hunt during the rainy season, has some conservation values as wildlife is not depleted by it.

5. *Community hunting by urban dwellers*

This is the category of licensed full-time professional or part-time hunters, who are bound by law to hunt certain categories and numbers of wildlife depending on their relative abundance. Most urban professional hunters hunt with rifles or single- or double-barrel shotguns. Licenses to hunt are issued by the game divisions of forestry departments in Cameroon, Nigeria, and Liberia and by the Department of Game and Wildlife in Ghana. Permits are not granted to hunt species that are prohibited or fully protected (denoted first schedule). Occasionally, permission may be given to hunt animals that are partially protected by law (second schedule). A list of these species is contained in the Nigeria "Wildlife Conservation Law," the Liberia "Wildlife Conservation Regulation," and the Ghana "Game Laws."

Market structure of bushmeat trade

"Market structure" is here taken to include such diverse economic phenomena as marketing channels and preservation methods, while "structure" is the organizational characteristic that determines the relationships between sellers in the market and the buyers. They are in fact the organizational characteristics that influence the nature of competition and prices of commodities within the market.

The production, distribution, and marketing of bushmeat in West Africa are not as complex as it is for some agricultural products. In most cases, there seems to be a general simple pattern in this part of Africa. The starting point of the distribution channel is the producers, in this case the hunters. Most producers reside mainly in rural areas, and the producer either sells his fresh product locally for consumption in rural areas or gives it to his wife first for preservation and subsequently to a village collector. The wife generally has two methods of preservation: soft smoking or smoke-drying before she sells directly to village collectors. The producer can also sell directly to a village collector who will carry out the smoking process himself. Alternatively, the producer's wife may sell fresh carcasses to roadside sellers who carry out the smoking exercise and sell by the roadside simultaneously. Otherwise, the roadside seller may have to buy from the village collector. The urban retailer can only buy from the village collector (often in smoke-dry condition) or from roadside sellers. The ultimate consumer buys mostly from the urban retailer in markets. Occasionally, an ultimate consumer buys from a roadside seller.

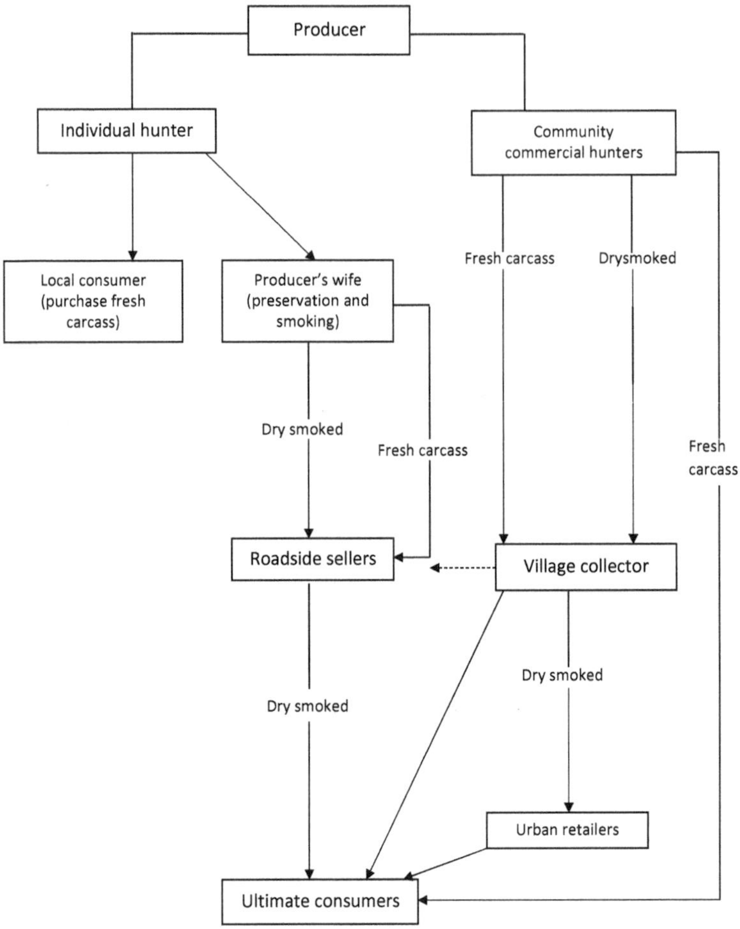

Fig. 1 Production and marketing structure of bushmeat in West Africa (Olawole and Ajayi, 1975).

Another aspect of the market structure is posed by traditional hunting tribes who may travel several kilometers and organize communal hunting for several weeks and eventually smoke-dry their products that are sold directly to village collectors and to urban retailers who eventually sell to ultimate consumers. They may also sell fresh carcasses directly to ultimate consumers in urban areas. Fig. 1 is an illustration of production and marketing structure of bushmeat in West Africa.

Preservation of bushmeat

Throughout West Africa, the only method used to preserve bushmeat is by smoking. Two types of smoking were identified: soft smoking and dry smoking. Smoking is usually carried out either in a mud-built smoking chamber or drum. Salting or biltong sun-drying is not common. Bushmeat that is dry-smoked can last 2–4 weeks, depending on how dry it is, and since the demand for it is so high, most of the commodity is sold before decomposition.

At present, there is probably no technologically similar alternative preservation method that can be suggested for bushmeat in West Africa. First, the tropical forest region is so humid that attempts at sun-drying meat for biltong leads to decomposition. Secondly, salting as method of preservation can make bushmeat more expensive and out of reach for common people. In fact, most people would prefer to utilize salt in a more economical way than for bushmeat preservation.

However, the way in which bushmeat is smoked in parts of West Africa leaves much to be desired. First, there is the question of hygiene. Smoking should be done in a manner that meets with the highest standards of sanitation. It is also desirable that dry smoking is affected by retaining the meat in a smoking chamber for as long as it is practicable, to reduce the moisture content to the barest minimum to ensure that it does not decompose quickly. Determination of minimum moisture content that would meet this requirement, however, remains a subject of investigation. Figs. 2–5 shows some some large mammals in Africa and their utilization.

Fig. 2 A male kob (*Kobus kob*)—a common antelope of the Guinea savanna grassland in West Africa. It is commonly utilized for food.

Fig. 3 A female waterbuck *K. defassa*—a common antelope of the West African forests and savanna grassland. It is one of the species that is usually utilized for food and by-products.

Fig. 4 Western hartebeest *Alcelaphus buselaphus* —a common antelope of the African forests and the savanna usually utilized for food and by-products.

Fig. 5 The African elephant *Loxodonta africana africana*—a common tourist attraction in African game reserves and national parks.

Economic importance of forest wildlife and its products in West Africa

West Africa has the highest human population density compared with other parts of Africa, south of the Sahara (Milligain and Ajayi, 1978). The human population continues to grow rapidly, and so are the natural accompaniments of their demands, mostly animal protein. Biologists and policy makers have been looking at wildlife as possible alternatives or as supplementary sources of animal protein. However, available information on the value of wildlife as a source of meat till date, especially in West Africa, has been almost entirely qualitative. The need, therefore, arises to collate existing information, scattered as it might be, to quantify the significance of this resource in the tropical forest regions of West Africa, where it is difficult to substitute wildlife with domestic animals, in large enough quantities to meet the demands of the population growth.

Work has been carried out on the importance of bushmeat in some parts of Ghana and Nigeria; however, nowhere in West Africa has any comprehensive country-wide survey been conducted with regards to the economic importance of bushmeat and animal products in either rural or national economies. In Ghana, Asibey (1970, 1974a,b, 1976, and 1978) has carried out market surveys in some selected urban and village markets to determine the economic importance of bushmeat, particularly the grass cutter *Thryonomys swinderianus* in the economic development in Ghana. In Nigeria, Charter (1970) has reported on the importance of wildlife in rural economies in the thickly populated south. The Federal Department of Forestry (1977) has also reported export figures of forest animals and by-products from Southern Nigeria, while FAO (1967) published reports on the value of some selected by-products exported from Sierra Leone and Ghana in 1962 and 1965, respectively. Apart from these references, there are no published current data available on the importance of wildlife in national and rural economies in Africa.

Bush meat in some national economies

In the absence of published current information on the type of comprehensive, country-wide survey referred to in the preceding text, efforts have been made to extract data from these references on the importance of forest mammals in national economies (Table 1). Table 1 also relates the percentage mean annual consumption of animal protein to human populations in five countries of West Africa. In all cases, it is evident that wildlife and particularly forest mammals account for 20%–90% of total animal protein consumed. Apart from the general economic importance of wildlife in the gross domestic products of the five West African countries on Table 1, there are no up-to-date empirical data of bushmeat and wildlife trade in Cameroon, Nigeria, and Ivory Coast. However, the latest information on Liberia and Ghana is presented here as mere indicators of the level of dependence of West Africans on wildlife. Liberia lies entirely on the tropical rain forest belt of West Africa, with an annual rainfall of above 80 inches over most of the country. The forest region is unsuitable for large-scale production of livestock, and hence, most of the Liberian population depends on wildlife for meat. In the counties of Grand Gedeh and Grand Bassa, wildlife is the only source of meat known to the people, and here, as in most parts of Liberia, wildlife is merely referred to as "meat" as there is no other substitute for it as a source of animal protein, except for a few domestic animals.

Bush meat in the economy of Ghana

From the list of wild animals eaten in Ghana, today, rodents, bats, anteaters, and primates form about 60% of the wild animal protein source, while the remaining 40% are birds and reptiles. Data collected in Accra on a year-long basis revealed a minimum consumption rate of 98,160.33 kg of bushmeat (mostly mammals), worth about US$15,514 (Asibey, 1978).

Table 1 Economic importance of forest wildlife in some West African countries.

Countries	Cameroon	Nigeria	Ghana	Ivory Cost	Liberia
% mean annual consumption of wild animal protein	78%–80% among rural people in the rain forest south	20% among rural people in the rain forest south	73% for the whole country	70% Southern part in the tropical rain forest	80%–90% for the entire country
Sources of information	Akum (1979)	Charter (1970), Ajayi (1978)	Asibey (1978)	Asibey (1978)	Ajayi (1979)

Asibey (1978) also reported that bushmeat contributed 44% and 31% of the protein consumed in Wiawso and Asempaneye districts of Southern Ghana, respectively. Fish contributed 35% and 31%, respectively. Hence, domestic sources of meat in this part of Ghana contributed 10% and 9%, respectively (Table 2). The foregoing estimates show the relatively high degree of dependence of West Africans on forest wildlife.

According to the Ghana "Wildlife Development Plan Volume 6 sustainable use of bushmeat 1998," the annual volume of bushmeat harvested by hunters in Ghana in 1996 was estimated at 384,991.8 tons worth about US$350million, while the total annual bushmeat consumption was estimated at 225,287 tons valued at US$205million. At the same time, marine and freshwater fish contributed 490,000 tons, while the total meat production from domestic animals (cattle, sheep, goats, and pigs) was 38,533.3 tons. The report of the wildlife surveys as contained in the Ghana Wildlife Development Plan stated that in terms of local markets, the volume of bushmeat traded annually in local markets is estimated at 91.731 tons valued at US$83,391million. This clearly shows that bushmeat contributes significantly to Ghana's national animal protein production and consumption. The total population of Ghana in 1998 was 12,296,081. For a country of this population, the contribution of bushmeat to the total protein consumption in Ghana and to its economy is significant.

Bushmeat in the economy of Liberia

Ajayi (1979) estimated that the value of bushmeat sold annually in Monrovia market was about $44 million. However, it was noted that out of the total bushmeat consignment that were meant for Monrovia in 1979, only 20% were actually sold in market places, while the rest were bargained from door to door by customers.

The estimate of annual production of bushmeat in Liberia was made by Ajayi (1979) at 105,000 tons; this volume, when the mean price for bushmeat in Liberia was applied, translated to about US$210m. However, the annual bushmeat consumption was estimated at US$66m by Ajayi (1979). The later estimate was about 4% of Liberia gross domestic product (GDP). Thus, the total national estimates for bushmeat at different period range from US$40 to US$210m. This estimate places bushmeat alone almost at per with timber in Liberia's national economy.

Bush meat survey of 1989–90 as food security index in Liberia

In 1989–90, the World Wildlife Fund and Forestry Development Authority (FDA) conducted a bushmeat survey in Liberia. The survey recorded about

Table 2 Species composition and numbers of animals killed by hunters in different zones of Ghana during 1-week survey.

Species common and scientific names (Ghana Wildlife Development Plan 1992)		Numbers killed							
Common name	**Scientific name**	**Assin Fosu**	**Goaso**	**Kumasi**	**Jasikan**	**Tumu**	**Total no.**	**Av. wt. (kg)**	**Total wt. (kg)**
Bushbuck	*Tragelaphus scriptus* Pallas 1766	–	–	–	–	–	9	41.7	375.3
Red river hog	*Potamochoerus porcus* (Linnaeus 1758)	–	–	–	–	–	4	85	340
Kob	*Kobus kob* (Erxleben 1777)	–	–	–	–	2	2	70	140
Oribi	*Ourebia ourebi* (Zimmermann 1783)	–	–	–	–	2	2	17	34
Bossman potto or potto	*Perodicticus potto* (Muller 1766)	–	–	1	1	-	2	1	2
Mona monkey	*Cercopithecus mona* (Zimmermann 1780)	–	–	1	–	–	1	6	6
Francolin	*Francolinus alboquolaris* or *F. bicalcaratus*	4	1	–	17	2	24	0.5	12
Total		98	49	62	251	22	482	–	2842.3

The average weights refer to fresh carcass weights and are based on average weights recorded in the market surveys.

35 species of mammals and reptiles in the commercial bushmeat trade, about half (15 species) of which are "highly threatened" and therefore classified as "protected animals" under the Liberian law. These species are high forest–dependent animals that are sensitive to excessive logging and vegetation change. Zebra duiker was dominant (about 66% of protected species) to Jentink's duiker (19%) and Diana monkey (5%). Forest primates such as the colobuses, Diana monkey, and chimpanzee that have a low resilience to hunting compared with duikers constitute about 3% of the trade in highly threatened species.

The survey was conducted in markets in large cities such as Monrovia and in rural areas, where the later had the highest varieties of animals. Animals regarded as crop pests such as black duiker, bushbuck, bushpig, cane rat, and porcupine constitute about 70% of total bushmeat sold in Monrovia markets and 30% in rural areas.

Bushmeat prices in Liberia

Prices of whole carcasses of bushmeat vary between an average of $40 for antelopes, $20 for primates, $57 for bushpig, and $5 for rodents. Irrespective of species, smoked-dry "quarters" of bay duiker, water chevrotain, and mangabey were sold in the range of $4–$7 each. While "quarters" of larger mammals such as Jentink's duiker, bushbuck were sold at $10–$14, and whole carcase of cane rat or porcupine also cost about $10–$14.

The information provided in this chapter was obtained between 1979 and 2009. It is therefore probably out of date in some details, but most of the facts expressed here represent the general picture of wildlife utilization in West Africa subregion today. The general exceptions are perhaps the declining status of wildlife and forest areas, which means there are less forest products including wildlife products and their derivatives such as bushmeat.

Prices of bushmeat per kilogram dry weight showed that primates ($2.15/kg) and antelopes ($1.49/kg) were the most expensive, while the main bushmeat price for Liberia in 1990 (all species considered) was $1.32/kg. In rural areas, prices of primates were twice the price of other meat items.

Inter-commodity price comparison of bushmeat and domestic livestock

One important index in assessing the economic importance of bushmeat in the economies of West African countries is the price comparison of this food item with other sources of animal protein, that is, domestic livestock. Market prices were abstracted either from published information or by direct market surveys. Countries for which data were available were Cameroon (Akum, 1979), Nigeria (Olawole and Ajayi, 1975), Ghana (Asibey, 1978), Ivory Coast (N'Guessan, personal comm.), and Liberia (Ajayi, 1979). The comparative commodity prices are shown in Tables 3–7.

There are some interesting facts that emerge from these tables. The first is that in almost all countries, bushmeat fetched higher prices than domestic animals.

Table 3 Intercommodity price comparison of bushmeat and domestic animals in Abidjan, Ivory Coast (1978).

Species wildlife	Price US$/kg
Grass cutter—*Thryonomys swinderianus*	10.00
Domestic animals:	
Beef	3.00
Pork	6.00
Mutton	3.50

Source: N'Guessan (personal comm.).

Table 4 Intercommodity comparison of bushmeat and domestic animals in Monrovia, Liberia, 1979.

	Status of carcass	Price US$/kg	
Species wildlife	Market 1 Rally time market	Market 2 General market at waterfront	Mean US$
Grass cutter—*Thryonomys swinderianus*	Fresh carcass	1.0 (Fresh ample)	1.0
Brush-tailed porcupine—*Atherurus africanus*	Fresh carcass	2.5 (Fresh ample)	2.5
Gray duiker—*Silvicapra grimmia*	4.0 (dry-smoked)		4.0
Red colobus—*Colobus badius*	3.0 (soft- smoked)		3.0
Mangabeys—*Cercocebus* spp.	3.0 (dry-smoked)		–
Spot nose monkey—*Cercopithecus ascenius*		2.5 (Fresh)	2.5
Domestic animals			
Beef	–	–	2.5
Pork	–	–	2.0
Mutton	–	–	4.0

Source: Ajayi, S.S., 1978. The utilization of Tropical Forest Wildlife: state of knowledge and research priorities. *Invited Special Paper Eight World Forestry Congress*, Jakarta Indonesia, 16–28 October, 1978, Ref. FFW/8-6:1–3.

Table 5 Intercommodity comparison of bushmeat and domestic animals, Ibadan, Nigeria in 1975 (summary for four markets).

Species wildlife	Price US$/kg
Hare—*Lepus cuniculus*	7.2
African giant rat—*Cricetomys gambianus*	5.0
Grass cutter—*Thryonomys swinderianus*	9.6
Tree tyrax—*Dendrohyrax arboreus*	6.4
Gray duiker—*Silvicapra grimmia*	6.8
Black duiker—*Cephalophus niger*	5.2
Bushbuck—*Tragelaphus scriptus*	5.2
Domestic animals	
Beef	4.2
Pork	4.4
Mutton	2.8

Source: Ajayi, S.S., 1975. Domestication of the African Giant Rat. Kwara State Publishing Co., Ilorin, Nigeria.

Table 6 Intercommodity price comparison of bushmeat and domestic animals in Sabouang, South Cameroon 1978.

Species wildlife	Price US$/kg
Hare—*Lepus* spp.	3.75
Brush-tailed porcupine—*Atherurus africanus*	4.375
Duiker—*Cephalophus* spp.	2.50
Monkey—*Cephalophus* spp	3.75
Baboon—*Papio anubis*	2.10
Civet—*Nadinia* spp.	1.875
Domestic animals	
Beef	2.50
Pork	2.00
Mutton	3.500

Source: Allo (personal comm.).

This can be attributed partly to production cost, that is, cost of weapons and hunting outfits, processing, and, more importantly, to transportation resulting from retail sales that tend to increase prices by as much as 25%–50% from the producer to ultimate consumer (O'Neil, personal comm.). The most important reason why bushmeat is more expensive is that the ever-rising demand for this rather diminishing resource, due to high human population pressure, inevitably resulting in the increase of prices. It is important to appreciate that bushmeat is highly favored by West Africans, and the little that reaches the markets is regarded as a delicacy whose price cannot be compared

Table 7 Intercommodity/price comparison of bushmeat and domestic animals, in Accra, Ghana in 1970.

Species wildlife	Price US$/kg
African giant rat—*Cricetomys gambianus*	0.51
Grass cutter—*Thryonomys swinderianus*	1.06
Brush-tailed porcupine—*Atherarus africanus*	1.06
Gray duiker—*Silvicapra grimmia*	0.88
Bay duiker—*Cephalophus dorsalis*	0.86
Blackbuck—*Cephalophus niger*	0.79
Bushbuck—*Tragelaphus scriptus*	0.66
Green monkey—*Cephalophus aeothiops*	0.79
Domestic animals	
Beef	0.30
Mutton	0.61

Source: Asibey, E.O.A., 1970. Shai, Hills Bush Meat Production Project. Unpublished MS. Department of Game and Wildlife, Accra, Ghana, pp. 21, 26, 28; Asibey, E.O.A., 1974. The grass cutter (*Thryonomys swinderianus Temminck*) T, Ghana. Symp. Zool. Soc. Lond. 34, 161–170; Asibey, E.O.A., 1974. Wildlife as a source of protein in Africa South of the Sahara. Biol. Conserv. 6(1), 33–39.

on the basis of weight for weight with domestic animals, but rather on the connotation of a "rare commodity" that is consumed in urban centers by the privileged few. It is right to say, however, that in spite of the high prices, demand for bushmeat is very high.

The second observation is that the species of forest mammals that were listed on Table 3–7 are mostly small mammals, that is, rodent, duikers, and monkeys. Large herbivores such as buffalo and elephants are completely absent from the list. From interviews with hunters and various authorities concerned with the management of wildlife in these countries, it was clear that the species listed in the succeeding text are the more prolific ones that survived the high hunting pressure of the last five decades or so. Large mammals have been virtually exterminated.

In most places, when the price of bushmeat was compared with other commodities, it showed that the cost of bushmeat and fish was similar. In Monrovia markets, duikers cost about $3.0/kg, chicken $3.00/kg, and beef $2.0/kg, and fish may cost as low as $1.0/kg. Domestic livestock meat was uncommon in the markets because they were rarely kept in rural areas and were regarded as economic securities for families that owned them.

Tables 3–7 present the general feature of interprice comparison of bushmeat and livestock in West Africa. It shows that bushmeat fetched higher prices than domestic animals. This was attributed to production costs: costs of weapons, hunting outfits, processing, and transportation, which tend to raise the prices of bushmeat items from rural areas to urban centers by as much as 5%–50%, that is, from the producer to the ultimate consumer.

Price fluctuations

Various fluctuations in prices of bushmeat were identified during the survey. There were price fluctuations within the same market. Intermarket price fluctuations and seasonal price fluctuations were reported in most places. With the exception of Liberia's "Rally Time" and "general market at waterfront," there were neither grades nor standards for selling these products. As this product was ungraded in Liberia as in most parts of West Africa, buyers often faced considerable inconvenience and uncertainty in price discrepancy. Although this aspect was identified as a deficiency in bushmeat trade, the prices quoted on Tables 3–7 were based on a method whereby an interviewer posed as a genuine buyer, to ensure good bargaining. The person who recorded his findings stood at a distance, and it was found that after a compromise on the price of a piece of a given species was reached, the piece was weighed. In this way, it was thought that the errors and biases that might arise from the ungraded products were minimized.

It can be concluded from the various studies on the economic importance of bushmeat in national economies, intercommunity price comparisons of bushmeat and domestic animals, that man's dependence on wildlife, especially in West Africa's tropical forest areas, is higher than expected. Also it becomes clear from Tables 3–7 that because of the increasing demand for bushmeat the wildlife resource is diminishing and consequently it is becoming dearer. All these facts point to the need for policy makers and development planners to give priority to wildlife management to ensure higher productivity. If the forest wildlife is exterminated, there is hardly any possibility of substituting it with domestic livestock in the near future because of the ecological and disease limitations that are inherent in the tropical forest ecosystem.

CHAPTER 3

A comparison of carcass composition and nutritive value of some West African forest mammals and domestic animals

Introduction

Although wildlife is widely utilized for food all over Africa, till date, there is little empirical information on their carcass quality and nutritive value as is available for domestic sources of meat. Talbot et al. (1965) have, however, shown that for game meat (elephant, Coke's hartebeest, wildebeest, and zebra) in East Africa, the nutritive value of the protein was high and the vitamin content much higher than that of beef, mutton, or pork from temperate zones. The authors also showed that there are striking differences between wild and domesticated stock. The killing-out percentages were presented as follows: in zebra steers in East Africa 40%–60%, wildebeest 50%, topi and Coke's hartebeest 53%, Thomson's gazelle 58%, impala 60%, and Grant's gazelle 63%. It was further demonstrated that unlike domestic animals, the killing-out percentage of wild herbivores is relatively constant within each species. They also found that game meat has a lower fat content and relatively higher lean meat all year round compared with domestic animals. Apart from Asibey (1978), Tewe and Ajayi (1977), Ajayi and Tewe (1978), and Tewe and Ajayi (1979), there is little published information on the nutritive quality of forest mammals in West Africa. Table 8 compares the percentage of dressed carcass values of pig, sheep, cattle, domestic rabbit, and some wild rodents (grass cutter and African giant rat). It shows that wild rodents compared favorably in carcass quality with rabbit and other domestic animals. However, as far as it is known, this is about the only available information on this subject, and therefore, further research is necessary into the carcass composition of West African mammals, before further conclusions can be drawn.

https://doi.org/10.1016/B978-0-12-816962-9.00003-X

Table 8 Live and carcass weights of some forest and domestic mammals.

Animal species	Domestic rabbit		Cattle	Pig	Sheep	Dwarf forest goat	Grass cutter	African giant rat	
Average live weight (kg)	1.569	1.851	643.6	93.9	27.1	15.67	5872.1	1.033	0.919
Average dressed carcass weight (kg)	0.792	0.948	249.4	70.2	13.9	7.94	3746.4	0.534	0.474
Average killing out (%)	50.0	51.4	38,8	74.8	49.8	50.59	63.8	51.5	51.6

Nutritive value of bushmeat

Table 10 compares nutritive value (fat, protein and ash, and carbohydrates) of West African forest mammals with cattle, sheep, and pig. In general, domestic animals are lower in protein and carbohydrate than their wild counterparts. For example, the range for carbohydrates is from 1.1% in red river hog to 6.0% in forest genet, whereas the range for domestic animals on similar environment is from 0.8% in pork and beef to 1.3% in mutton. Likewise, the range in protein content is from 16.1% in tree hyrax to 55.4% in forest genet. For domestic animals, the picture is that of lower values: from 11.25% in pork to 19.6% in beef. It can, therefore, be concluded from available data on the proximate compositions of 12 forest mammal species and three domestic animals that wildlife is superior in terms of protein and carbohydrate contents. In a separate study, Asibey (1978) also found that the protein content of 12 forest mammals (Table 9) range from 21.7% in porcupine to 24.9% in bushbuck and concluded that they compare favorably with domestic animals. Table 10 shows the proximate composition and mineral contents of some forest mammals and domestic animals. However, it is important to note that very little work has so far been done on the mineral content of African forest mammals and virtually nothing on the amino-acid composition of the proteins (Table 10).

Abundance and trend of forest mammals in West Africa

Although fragments of information exist for various species utilized for food and animal products and for their economic importance, however, as far as it is known, no country had a record of the status and trend of animal

Table 9 Protein content of some forest mammals (Asibey, 1978).

Species	Protein content (%)
Brush-tailed porcupine—*Atherurus africanus*	21.7
Grass cutter—*Thryonomys swinderianus*	22.7
African giant rat—*Cricetomys gambianus*	22.45
Bay duiker—*Cephalophus dorsalis*	24.1
Black duiker—*Cephalophus niger*	23.9
Maxwell's duiker—*Cephalophus maxwelli*	21.8
Bushbuck—*Tragelaphus scriptus*	24.9
Palm civet—*Nandinia binotata*	22.8
African Civet cat—*Civettictis civetta*	23.2
Spot-nosed monkey—*Cercopthecus petaurista*	24.3
Mona monkey—*Cercopithecus mona*	24.5
Colobus monkey—*Colobus polykomos*	24.0

Table 10 Proximate composition and mineral contents of some forest mammals and domestic animals (Ajayi, 1979).

Forest mammals	Moisture G.	Protein G.	Fat G.	Ash G.
Rodents				
African giant rat—*Crycetomys gambianus*	65.99	18.99	11.0	3.9
Grass cutter—*Thryonomys swinderianus*	54.8	20.7	0.7	0.5
	77.4[a]	26.2[a]	7.4[a]	1.0[a]
	69.7[b]	18.8[b]	8.9[b]	1.2[b]
Tree hyrax	78.7	16.1	2.9	1.9
Primates				
Green monkey—*Cercopithecus aethiops*	80.3	17.6	1.3	1.1
Proboscidea				
Elephant (skin)—*Loxodonta Africana cyclotis*	49.4	29.2	19.7	1.7
Carnivora				
Long-snouted mongoose—*Herpestes naso*	72.7	18.8	1.9	3.3
Forest genet—*Geneta opartuina*	31.8	5.4	9.3	6.0

Suidae				
Red river hog—*Potamochoerus aethiops*	7.1	23.8	1.6	1.1
	72.4[a]	27.2[a]	2.7[a]	1.1[a]
Artiodactyl	45.1	40.6	10.7	3.3
Royal antelope—*Neotragus pygmaeus*	74.5	23.4	0.9	1.2
Gray duiker—*Sylvicapra grimmia*	59.5	33.4	2.0	4.0
Bushbuck—*Tragelaphus scriptus*	47.6	50.9	12.2	3.7
Ox (beef)	73.8	19.6	12.0	1.0
	54.7[b,c]	16.5[b,c]	28.0[b,c]	0.8[b,c]
Sheep (mutton)	78.5	17.2	2.9	0.8
	62.5[b]	16.8[b,c]	19.4[b,c]	1.3[b,c]
Pig (pork)	54.8	19.4	13.4	0.8
	41.1[b,c]	1.2[b,c]	49.0[b,c]	0.6[b,c]

[a]Minerals mg/100 g of fresh edible tissue (Ajayi and Tewe, 1979a,b).
[b]In terms of grams/100 g (Ajayi and Tewe, 1978).
[c]Proximate values are expressed (Kramlich et al., 1973).

populations in their tropical forests. Consequently, till date, nobody knows what remains of forest mammals in West Africa. It is thus impossible to state precisely the population trend of wildlife in this part of Africa.

From some isolated studies, the following information, however, provides indications of population trends:

(a) Milligan et al. (1978) compared the population densities of 16 herbivores in 11 national parks located in savanna regions in seven West and Central African countries. The biomass densities indicated a considerable range from 214 to 4.043 kg/km^2. Using rainfall data for these areas to estimate primary production and potential carrying capacity (Coe et al., 1976; De Leeuw, 1977), there was an indication that the present biomass density levels of the West and Central African national parks were well below the potential of their habitats. In many cases, the biomass densities were only 10% of the potential carrying capacities. The author attributed the low biomass to overhunting and the lack of enforcement of game laws.

(b) The second indication used was the available export figures for forest mammals and by-products from Southern Nigeria in 1965/66 and 1966/67 financial year. There was a decline of about 80% in revenue that could be attributed to legislative control (because none existed) but to the rapid decline in the resource itself.

(c) The third indication came from the wildlife and national park legislations of the countries visited. Animals were usually classified into three schedules according to the Organization of African Unity (OAU) (Now African Union, AU) Wildlife Algiers Convention of 1968. In accordance with the convention, provision should be made under national legislations for lists of animals that are endangered and are, therefore, to be strictly protected (Schedule I) and of animals that are partially protected (Schedule II), that is, animals that can only be hunted for exceptional reasons (Schedule III). The latter includes unprotected animals that can be hunted under license. Although some of the legislations used were old, it is important to note the rather staggering list of species that are endangered today and are not strictly protected by law. Since legislations are old and apparently no surveys have been conducted to determine the status of listed species, it seems reasonable to combine Schedules I and II for each country and treat this as a current list of endangered forest mammals in West Africa. Even under the present classification, important species from the point of view of food and wildlife products are in danger of extinction. In Liberia, Ghana, and Nigeria, practically, all primates are endangered. Likewise, forest antelopes such as the bushbuck and all species of duikers, which are highly productive, are so intensively exploited that they are placed in Schedule I of many countries. The big carnivores (leopard and lion) whose skins are highly priced in the export market are also facing extinction in West Africa.

(d) Apart from the declining volume of trade in bushmeat in markets, recent market surveys in Nigeria indicate that only species that are highly productive and are, therefore, difficult to hunt out—such as rodents (Giant rat, *Cricetomys gambianus*, and grass cutter, *Thryonomys swinderianus*) form the bulk of the bushmeat trade. Except for Liberia, antelopes and monkeys are rarely sold because they are seriously depleted outside game reserves.

The foregoing account provides an indication of the decline in the status of forest mammals in West Africa. This decline could be attributed to overhunting, which is a direct result of higher demand for bushmeat—a phenomenon that is a natural accompaniment of the rising human population and a clear preference for bushmeat over other sources of animal protein.

It is important, however, to point out certain anomalies in the wildlife legislations of some West African countries:

Firstly, in Cameroon, Ghana, and Ivory Coast, the legislations were old and need to be updated. In Nigeria, no national legislation is yet operative, despite the autonomy granted for the administration of wildlife through the creation of the National Park Service. Furthermore, in Nigeria, there is no national policy for wildlife conservation despite the broad mandate given to the National Park Service.

Secondly, certain categories of animals, notably the large carnivores such as leopard and lion and elephant, are not adequately protected by law in several West African countries. Indeed, Liberia and Ivory Coast made no provision for the protection of lion and leopard, species that are said to be in imminent danger of extinction in both countries. Although elephant is to be listed as a protected animal for Nigeria, the draft law only makes provision for the protection of immature animals. It is logical to protect the productive older age groups and immature elephants, if the objective is to save this species from extermination.

Thirdly, Ivory Coast has banned hunting since February 20, 1974, and therefore, theoretically, all forest species are under legal protection. However, it was understood that in practice, this type of legislative measure could not be enforced as illegal hunting continued in most places.

CHAPTER 4

Animal products in traditional medicine

Wildlife is known to be an important aspect of the culture of West African tribes. Medicinal uses of a wide variety of forest animal products show the importance of this resource in the evolution of culture in West Africa. The general impression is that the inhabitants of the tropical forest zone are well advanced in medicinal uses of the flora and fauna of their environment for preventive and curative purposes. However, many species of forest animals are on the verge of extinction for this reason, and paradoxically, little research has been conducted in the field of utilization of wildlife for traditional medicine.

The method employed by the author consisted of questionnaires on the uses of a wide variety of game animals in the tropical forests, including reptiles, birds, and mammals. It was recognized that West Africans are rather secretive when it comes to giving information on ingredients that are used to cure an illness or to perform certain feats in witchcraft (in most cases native "doctor" will only release this information to the child he loves most within his family). Therefore, children of herbalists and native doctors were carefully selected to carry out the study. The results on the uses of forest mammals for witchcraft and medicinal purposes are presented in Table 11.

Wildlife Conservation in Africa
https://doi.org/10.1016/B978-0-12-816962-9.00004-1

Table 11 Utilization of tropical forest mammals for traditional medicine and witchcraft.

Order	Species	Part of body used	Medicinal uses
Primates	Gorilla—*Gorilla gorilla*	1. Bones	(a) An important ingredient in preparation of concoction for young children to drink for protection against diseases and witchcraft
		2. Left hand	(b) Also used to prepare witchcraft concoction to prolong pregnancy in women—an act of punishment for an enemy Commonly used to prepare concoction with herbs for children to drink for immunity against diseases, epidemics, and witchcraft
	Chimpanzee—*Pan troglodytes*	1. Skin 2. Bones and skull	Used in witchcraft prepared with other ingredients to a padlock. The user dictates his wishes against his enemy to the padlock and locks it. As long as the padlock remains in locked position, so will his wishes affect the enemy An important ingredient in a concoction for children to give them immunity against diseases and witchcraft
	Various monkeys	Skull	Used in combination with other ingredients to cure stomach ailments in children 6–12 months old
Proboscidea	Elephant—*Loxodonta africana cyclotis*	1. Meat 2. Bones with some meat attached	Used in preparation of medicines to be eaten by children for immunity against diseases Used with black native soap for bathing for good luck

Table 11 Utilization of tropical forest mammals for traditional medicine and witchcraft.—cont'd

Order	Species	Part of body used	Medicinal uses
Carnivora	African civet—*Civettictis civetta*	Anus	Used in traditional medicine for preventing convulsion in children
	Leopard—*Panthera pardus*	1. Bones 2. Skin	Mainly used in witchcraft to effect protection against the wishes of enemies (a) In high demand for witchcraft to effect a respectful personality for an individual
		3. Feces	(b) Also used in witchcraft whereby any weapon made of iron cannot penetrate the skin of the user. Commonly used in time of intertribal wars
	Lion—*Patnthera leo*	Heart	An ingredient in witchcraft preparations to track down thieves Used in witchcraft by warriors for courage
Artiodactyl	Bushpig—*Potamochoerus porcus*	1. Bone	Used in concoctions for children to prevent fever and associated illness when growing milk teeth
		2. Head	Used in traditional medicines with black soap for bathing for good luck. Mainly used by traders and businessmen
	Bushbuck—*Tragelaphus scriptus*	1. Horn	Used in witchcraft for inflicting bad luck against an enemy
		2. Bones	In preparation of concoctions with native black soap for bathing children 3 months–3 years. This is supposed to impact immunity against "all diseases"

Continued

Table 11 Utilization of tropical forest mammals for traditional medicine and witchcraft.—cont'd

Order	Species	Part of body used	Medicinal uses
Insectivora	Hedgehog—*Atelerix albiventris*	Whole body	Used in witchcraft for protection against witches
	Elephant shrew—*Rhynchocyon* spp. and *Petrodromus*	Whole body	Used to prepare a concoction with palm kernel oil and given to children in small doses to stop convulsion
	Pangolin *Manis (Smutsia) gigantea*	Spine	Used to prepare medicines for pregnant women to effect delivery of their young
Rodentia	Brush-tailed porcupine—*Atherurus* spp.	Spine	Also used in concoction for pregnant women to effect easy delivery of their young
	Crested porcupine—*Hystrix* spp.	Spine	Used in concoction to prevent miscarriage in pregnant women
	Flying squirrel—*Anomalurops beecrofti*	Tail	An important ingredient in traditional medicine for pregnant women to prevent miscarriage and to ensure that the child is eventually born healthy
	Grass cutter—*Thryonomys swinderianus*	Hair, skin, and intestines	Used in concoction for healing wounds Mixed and ground with black soap to bathe children to impart immunity on them against diseases
	Forest African giant rat—*Cricetomys emini*	Head	Use in witchcraft to win the love of a woman

CHAPTER 5

Utilization of tropical forests and the wildlife conservation in West Africa

Wildlife is an important component of the forest ecosystem and is dependent on the vegetation in order to thrive. Hence, it would not be adequate to consider the status and trend of tropical forest mammals without giving some consideration to their habitats.

Until the beginning of this century, tropical forest wildlife was reasonably undisturbed and was considered to be in balance with traditional human activities, mainly shifting cultivation and hunting with primitive tools such as snares, traps, and flint guns (Asibey, 1976). Toward the end of the 19th century and more pronounced during the first decade of the twentieth century, plantation agriculture, particularly for cocoa, oil palm, and rubber, was introduced to West Africa. This marked the beginning of large-scale utilization and decimation of tropical forest ecosystems. From that period onward, timber extraction for export market has gradually grown to a point that it constitutes a threat to the existence of tropical forests. It is pertinent to note here that the major factor that is responsible for the degradation of the tropical forest is economic, since rubber, cocoa, and oil palm have been and will continue to be important to foreign exchange earners for many countries in West Africa. Unfortunately, there was no early recognition of fauna conservation as a form of land use. When at some later stage steps were taken to set aside certain areas as game reserves, those early game reserves were not properly managed. Consequently, with the increasing population pressure and the high demand for meat, combined with drastic modification of the forest ecosystem, and for agriculture, wildlife was greatly reduced.

The forest ecosystem is complex. Thus, it makes room for many niches and hence a much higher diversity of highly specialized animals than in savanna grassland. What this means is that the specialized nature of some forest animals does not enable them to survive once the primary forest is drastically modified or destroyed. For example, the gorilla, the chimpanzee, and

Wildlife Conservation in Africa
https://doi.org/10.1016/B978-0-12-816962-9.00005-3

the bongo, to cite a few, can hardly survive in secondary forests. A study of the food habits of red colobus monkey showed that 51 items of its diet came from economic trees of primary forest. Red colobus monkey has therefore been restricted to very few undisturbed forest areas of West Africa for this reason. Once the remaining forest habitat is modified, the red colobus will be in danger of extinction in West African subregion.

The following is an account of the status and trend of tropical rain forest in some West African countries. Table 12 gives a summary of estimates of the total areas covered by tropical rain forests and those set aside as forest reserves (Persson, 1974).

Starting from Cameroon, the whole southern part of West African countries is covered by tropical rain forest, that is, a total of 17.5 million hectares. Out of this, approximately 100,000 ha is mangrove forest. There is no information available on the areas set aside as forest reserves. There are approximately 0.25 million hectares of tropical rain forest in the Republic of Benin, and this accounts for about 2% of the country's total land area.

Out of the 8 million hectares of tropical rain forest in Ghana, 1.8 million hectares have been largely constituted as forest reserves. From the areas of unreserved tropical forests, about 50,000 ha disappears every year. About 50% of timber exploited comes from forest outside forest reserves. It is, however, interesting to note that out of 8 million hectares of Ghana's tropical forest, about 0.6% has been committed to wildlife conservation (Asibey, 1976).

In the Ivory Coast, about 16 million hectares of tropical forest land existed out of which 3 million hectares disappeared between 1956 and 1966. This degradation according to Persson (1974) continues unabated and may lead to a serious depletion of high forests within the next few years. At present, 2.5 million hectares of closed high forests exist in Liberia out of which 1.6 is declared forest reserve.

Table 12 Areas of tropical rain forests left unutilized in some West African countries.

Country	Area of forest left (million hectares) (1979)	Forest classified reserved areas (million hectares)
Cameroon	17.5	–
Nigeria	31.0	1.9
Benin	2.1	–
Ghana	8.0	1.8
Ivory Coast	16.0	–
Liberia	2.5	1.6

The foregoing account of the extent and trend of West African tropical forest habitat forecasts a bleak future for the wildlife. Some urgent action is needed to arrest the rate of decimation of the tropical forests. If not, it will have serious repercussions for primary forest wildlife species in this part of Africa.

Some options in the management of wildlife on farmlands

The agricultural system of shifting cultivation, which has been practiced for centuries in most of the tropics, is still prevalent in West Africa. Under this system, the forest is cleared, and food crops are planted.

The secondary regrowth, together with food crops, provides a good habitat for certain wildlife species such as rodents and game birds. These animals are often hunted by farmers and provide a steady source of animal protein. Even though the "small game" on farmlands has long been recognized as a source of food, they have hitherto often been regarded merely as "pests" that should be eliminated. There have been no suggestions as to how wildlife could be managed on farmlands to maximize production while at the same time ensuring minimum damage to crops.

During the study tour of West Africa, the impression gathered was that farmland animals provide a significant portion of animal protein for farmers and others in rural areas, and their active management may well provide a complementary element to the present management practices. The suggestion to promote wildlife management on farmlands is a new concept to West Africa, which, like agroforestry, is a suggestion that could easily be discredited if it is not backed by intensive research findings and careful but flexible planning.

These and many other questions remain to be answered before such a management practice can be introduced extensively. However, this appears to be a vast but potentially rewarding field of research and could be attractive to researchers and planners alike. The results of this practice may well revolutionize animal production in tropical forests.

Possibility for introduction and reintroduction of species

"Introduction" is here defined as an intentional movement of an animal or plant from its biogeographical zone to an area where an ecological niche exists for the species but where the species has never been known to thrive before. "Reintroduction," on the other hand, applies to movement of animals from one to a similar biogeographical zone, which formed part of their former range.

The foregoing account of the status and trend of forest mammals in Africa shows clearly that a large number of species are in danger of local extinction. It will therefore not be out of place to consider the ecological implication of extermination of these species and the possibility of introducing and/or reintroducing certain species of wildlife. Extermination of certain species can and has led to undesirable ecological inbalance in some parts of the world.

First, extermination means the complete loss of an animal species along with its existing or potential economic resource and cultural values. A complete loss of a species may affect other species, thus disturbing the equilibrium of the ecosystem. Jungius (unpublished MS) cited an example of extermination of roe deer and red deer in their Abruzzo range in Italy, which led the natural predators, wolves and bears, to prey more heavily on domestic stock and *Abruzzo chamois*—which is now an endangered species.

However, one ecological implication of reintroduction must be critically examined. Species extinct from certain areas may not be successfully established if reintroduced, because human activities might have modified the original habitat to such an extent that "new" niches have been created; an example is artificial afforestation using exotic tree species in many parts of West Africa. This is perhaps the crossroad between reintroduction and introduction. This is because the new niches created cannot be allowed to be "vacant" indefinitely if the new niches are similar to those in other habitats elsewhere. On this basis, the principle of introduction is considered in some quarters to be scientifically justifiable.

However, it must be remembered that there have been many introductions, especially early in the 20th century, and some have been successful, while some were not. The following are some ecological implications of introduction:

Introduction of species could cause ecological imbalance and displacement of native species whose requirements overlap with the newcomer. For example, red deer introduced into Chile and Argentina was reported to be competing with the guemal and pudu for food and space, thus forcing native species to less attractive habitats (Jungius, unpublished). Sika deer introduced from Japan to Germany are competing with red deer and fallow deer (Jungius, unpublished).

Introduction could lead to population explosion and habitat destruction; for example, red deer introduction into New Zealand lacked natural control mechanisms and insufficient management resulting in population explosion and habitat damage.

Child (1979) also stressed the veterinary implications of introducing new species since it is also possible to import their diseases with them. Diseases from introduced species might have a devastating effect on the native fauna since the latter have not been previously exposed to these diseases and, therefore, have not built up a natural immunity against them.

The above outlines the general environmental implications of the conduct of reintroducing or introducing new species to West African tropical forests. However, since reintroduction is less controversial and appears to be feasible in West Africa, future consideration in this regard must take into account some practical prerequisites to this exercise:

Conduction of feasibility studies using the following guidelines:

1. Collect data on the ecology and behavior of the species to be introduced.
2. Reasons for its extermination.
3. A survey of the proposed reintroduction site to ensure that it would meet the ecological requirements of the species to be reintroduced.
4. To be certain that the reasons for the species extermination are no more in existence.
5. A provision of adequate protection and management for the reintroduction area. This should include veterinary supervision of the area.
6. Prior to release, animals must be held temporarily in the reintroduction area in order that they might be acclimatized and also to enable them to establish social groupings. This is in particular reference to gregarious species. This will ensure successful breeding when they are released.
7. Since it is necessary to follow up the performance of the animals in the reintroduced area, it is important to mark them or fit individuals with radio collars to enable them to be monitored.
8. Stocks from overseas must be examined to ensure that they are clear of diseases.

In conclusion, the concept of introduction or reintroduction of species should not be rejected as a management tool for reestablishing the fauna of West African tropical forests. It is important, however, to know the ecological implications of this concept and to follow the general guidelines stated above in order to avoid ecological disasters that have rendered some similar programs unsuccessful in the past.

CHAPTER 6

State of knowledge on tropical forest mammals and priorities for research

An overview of wildlife research projects

In order to identify gaps in knowledge on the management and utilization of forest mammals, a survey of past and ongoing research projects in this field was compiled for the six countries visited in West Africa.

Apart from the study of the biology of small mammals of economic importance such as the African giant rat, grass cutter, and bats and some isolated surveys on bushmeat marketing, other aspects of wildlife utilization seem to be neglected. On the basis of the current research projects, gaps were identified, and future research priorities and recommendations for the management and utilization of forest mammals were made.

Priorities for research

Through the course of study on the utilization of forest wildlife for food and animal products, by literature surveys, in consultation with appropriate officers of the forestry departments, and also by reviewing recent research projects on this subject, it became evident that there was a conspicuous lack of quantitative information on the value of wildlife in the socioeconomics of West African countries.

In the case of planning for development, where wildlife has to compete with other sectors of the economy for funds, the relevance of quantitative information on the importance of wildlife cannot be overemphasized. Therefore, it is suggested that surveys be carried out using the following guidelines in order to avoid biases and errors that may eventually render the

Wildlife Conservation in Africa
https://doi.org/10.1016/B978-0-12-816962-9.00006-5

final product worthless; the survey should be based on questionnaires that should have the following components:

1. General information on the individuals—name, village, sex, age, income, and size of whether or not hunting is a major occupation of the respondent
2. Production and processing, pattern of hunting, etc.
3. Preservation and transportation
4. Mode of processing
5. End use of the kill
6. Marketing

The following sampling techniques are appropriate for both rural and urban areas of the country to be surveyed. In order to have an appraisal of bushmeat production and marketing and to be able to trace bushmeat from the producer to the ultimate consumer, such a survey should be conducted as follows:

Both rural and urban areas should be covered. There should also be a household survey, and the sampling method should be "1 in K" type of sampling. By "1 in K" type of sampling, it is meant that the survey should be random in nature, with only one "household" to be interviewed out of every given number, say one out of 10 or 20 or 50 depending on the size of village or town to be sampled. This method of course will have as a prerequisite knowledge of the lists of houses in a sample area.

The "1 in K" type of sampling method implies that if one out of every 10 households were to be sampled, the selection of the first house is critical to the rest, that is, if you mix the addresses of the first 10 houses and choose one at random and that one happens to be number 1, the next must be number 11, 21, 31, and so on. If it is number 6, the others must be 16, 26, 36, and so on.

It is again necessary to define the word "household" in a survey exercise such as this, because in African tradition, there may be more than one family occupying the same house and even in urban areas where many families occupy a block of flats that constitute a "house"; the need for such a definition is more necessary. For convenience sake therefore, the term "household" is taken to mean those eating from the same plot. The head of the "household" is usually the one to be interviewed. Such a survey would eliminate biases and errors as far as possible, and the particular samples or respondents depending on the information sought can be classified into different categories subject by subject or on the basis of the general information given.

With the exception of Ivory Coast, Nigeria, and Ghana, there is no country in West Africa that had information on animal population and habitat monitoring. It is suggested that these studies should be accorded a priority since no effective development plan can take place without the knowledge of what remains of the wildlife, their trends, and interactions.

Everywhere in West Africa, wildlife is widely regarded as pests on farmlands to agricultural crops. From available information, these animals are the main sources of animal protein to the farmers in rural areas and the main source of cash to hunters. Hunting farmland animals is also regarded as a means of sport in Ghana and Cameroon. The species concerned such as the grass cutters, bush fowl, giant rats, squirrels, and fruit bats are relatively prolific and are adaptable to the presently modified tropical forests. It is desirable to adopt a realistic approach to the management of farmland wildlife and examine how to harmonize the management of these animals and food crops in agriculture, rather than regard wildlife merely as pests to be exterminated. It is important to study the methods of management of grass cutters, bush fowl, squirrels, and giant rats on cassava, rice, and maize farms and grass cutters in sugar cane plantations so that man can begin to crop plants and animals on the same land on systematic rational basis, thus putting the forest lands on a more intensive use. This is a system that would be similar to the present practice of agroforestry. It is even possible to crop, on a rational basis, some other forest animals such as fruit bats, which are currently utilized for food but are regarded as a nuisance in forest areas. However, in order to embark on such a scheme, it would be necessary to have information on their productivity, behavior and patterns of movement and migration, and their effects on forest trees.

In order to be able to harmonize the management of wildlife with agricultural crops, the following investigations should be made, and the results to be used as a basis of its management scheme:

a. The kind and extent of damage done by a wildlife species to a particular crop and the period of this damage are done.
b. At what stage of development of that crop is this damage done?
c. At what stage can the damage be considered economically intolerable?
d. What is the reproductive cycle and reproductive performance of the animal species concerned?
e. Is there any correlation between population growth and reproductive cycle and intensity of damage to crops?
f. From the knowledge of fecundity, mortality pattern, sex ratio, and age structure of animals, what is the percentage of animals for each sex and

age group that can be cropped in order to reduce the population to a level in which the damage it does to crops is minimized? What is the optimum rotational and seasonal cropping scheme that could be evolved for efficient wildlife cropping that would ensure a sustained yield of animal protein?

Modern legislation that would ensure efficient utilization of forest wildlife is mostly lacking in West Africa. Under the present exploitive practices in Liberia and Nigeria, where there is virtually no legislation restraint to the exploitation of wildlife, forest wildlife may well have a chance of survival. Modern legislation for each country would be based on the inventory of species and a sound knowledge of population trends, especially those of economic importance, their relative abundance, and population trends that would allow categorization of forest fauna. This classification will show which species are threatened and should be prohibited in terms of exploitation and those that are less rare and could be partially protected and exploited on exceptional grounds. It will also enable categorization of species that could be hunted. Furthermore, such legislation can prohibit export of animals from West Africa for zoos and trade in certain wildlife products such as ivory, hides, and skins. In this regard, the use of wildlife in forest national parks, especially in Ivory Coast for tourism, whereby animals are specially protected for the sole enjoyment of tourist, again underlines the importance of the need to develop forest wildlife as a source of protein to the people.

The principle of "thou shall not touch" and "allow nature to regulate itself" as a policy for the management of parks and equivalent reserves is in many situations valid. In buffer zones and utilization areas, requirements in terms of management for animal protein production are widely different. Secondly, it is well known that most species of animals can withstand a certain amount of "culling" per year. An advantage can be taken of this to "take off" this percentage, with modern management practices that are now available, to ensure a sustained yield. Hence it is possible that utilization of forest wildlife can be conducted simultaneously for game viewing and game cropping if a careful ecologically based management plan is made for appropriate game reserves. An obvious advantage of this practice, apart from providing valuable animal protein to rural people, is that it can reduce illegal hunting, since local hunters would participate in game cropping.

The highly specialized and peculiar nature of some forest animals have made it impossible for them to survive outside their habitats, especially where tropical forests have been modified by man. Thus, a large number of primates, for example, gorilla, colobus, and artiodactyls such as the bongo and yellow-backed duikers, are on the verge of extinction in West Africa. Rehabilitation of these and other species depends on detailed knowledge of their habitat requirements and the degree of their specialization, a knowledge of their general biology, food habits, reproductive performance, etc.

An area of wildlife utilization, which has hitherto not been recognized as of importance to the inhabitants of the forest regions of West Africa and which has contributed in no small way to the declining trend of forest wildlife, is the use of forest animals and their products for medicinal purposes. This is an area where more information is required on authentic uses of wildlife in curative medicine and in which cases animals can be put to more efficient use and which areas of utilization of wildlife for medicinal purposes are merely destructive and should be discouraged.

There is still a big gap in our knowledge of the nutrient quality of bushmeat in West Africa. Apart from Asibey and Eyeson (1975), Tewe and Ajayi (1977), and Ajayi and Tewe (1978), there are virtually no known publications on this topic. These articles merely give the live and carcass weights proximate compositions (crude fiber, protein, and fat) of some species of mammals. However, there are several forest species that are of importance as food items that the earlier studies did not cover. Furthermore, detailed studies are necessary to uncover their mineral composition, the nature of their fatty acids, and the composition of their amino acids in order to provide a basis for comparison with their domestic counterparts.

Already there are statements, based on hard facts, that East African wild ungulates contain more lean meat than their domestic counterparts. There are also claims that wildlife in general contains less saturated fatty acids than their domestic counterparts (saturated fatty acids are known to be inimical to good health). But these claims have not been well substantiated in West Africa.

Summary and recommendations

The study on utilization of forest mammals and their by-products in West Africa has shown that practically, all species are utilized for food and a wide variety for ornamental, medicinal, and scientific purposes. It also shows that forest mammals have contributed in no small way to the development of zoos all over the world. This chapter reveals that both primitive and modern tools are employed side by side to exploit forest mammals. The various methods of exploitation that were examined showed that commercial community hunting was indiscriminate in nature and most wasteful and should be discouraged. The method of preserving bushmeat by dry smoking, if done in a hygienic way, was perhaps the most practical and inexpensive method of getting bushmeat from rural to urban areas. Salting as an alternative method is considered too expensive for the ordinary man to afford, and sundrying to make biltong is not as effective as smoking because meat subjected to this method of preservation decomposes too quickly in the humid tropical environment of West Africa.

Various studies on the economic importance of bushmeat and animal products revealed rather impressive quantitative information that shows how heavily West African countries are dependent on wildlife. It shows that wildlife constitutes as much as 20%–90% of the animal protein in six countries of West Africa and a comparison of intercommunity prices showed that generally, bustmeat fetched higher prices than domestic animals. Thus, information contained in this book on this subject could help policy makers and planners to refocus their attention on the management of these policy makers and planners to refocus their attention and place more emphasis on the management of this resource.

At present, there is little information on the nutritive value of forest mammals that would allow for a comprehensive comparison with that of domestic animals. This is a field of research for nutritionists and biochemists in order to fill the gap in knowledge on the quality of game animals.

As far as it is known, there are not many ongoing wildlife population monitoring program for tropical forest mammals in this part of Africa. The indexes of decrease in trade figures, ratio of animal biomass and the potential carrying capacity of wildlife habitats, and the long list of endangered species as contained in the wildlife legislations of several countries only provide an indication of a rather sharp and accelerating downward trend of tropical forest wildlife from the beginning of the 20th century.

Likewise, recent information on the extent, mode, and rate of modification of tropical forest vegetation shows that wildlife habitats are being drastically modified and degraded at such a rapid rate that if nothing is done quickly to arrest the situation, tropical forest wildlife has very little chance of survival in West Africa.

An examination of West African fauna and agricultural practices shows that rodents, bush fowls, guinea fowls, and fruit bats can possibly be managed compatibly with agricultural crops as a form of multiple land use. West African rodents, such as grass cutter and the African giant rat, which are highly productive and well favored as sources of meat, have often hitherto been regarded as pests to various farm crops. However, they can be managed in such a way that wildlife and farm crops will be harvested economically from the same land.

Also, the various forest plantations of both indigenous and exotic tree species can well provide valuable niches for reintroduced or, in exceptional circumstances, possibly, introduced species of wildlife and, if both are managed together, may constitute a new approach to the conservation of forest wildlife in West Africa. The proposed multiple land use may thus lead to increased production of animal protein. These proposals are areas for future research. It must be pointed out, however, that apart from the research prerequisites, the management of agriculture with rodents, like the taungya system, is a sensitive issue that can be easily discredited if not backed by careful and rather flexible planning and execution.

Wildlife legislation in several West African countries has many inadequacies. Most laws are old and need revision. Furthermore, some legislations contain several anomalies and do not give adequate protection to species of economic importance that are now in danger of local extinction. In this connection, it is relevant to point out that the conservation method whereby state governments ban hunting completely is considered impractical and a negative approach to wildlife management.

A more positive approach will be to strengthen wildlife management services in terms of personnel, equipment, transport, and other facilities and to give wildlife divisions a certain amount of autonomy and to cut through red-tapeism and bureaucracy. This will enable field wildlife managers to attend to urgent and irregular conservation issues.

Education of citizens on the value of wildlife management is of course an essential tool in wildlife management, because no government project can succeed without the goodwill of the people. It is also important that local people, that is, those that are closest to national parks and equivalent reserves, be involved actively in the running of these conservation areas.

It must be noted that most of the existing game reserves in West Africa were once traditional hunting grounds of the local inhabitants. If any government were to reserve such lands by law, it is only realistic to involve their participation and also give them part of the benefits that accrue from the utilization of wildlife resources. This can provide employment opportunities especially for the unskilled section of the rural population.

Lastly, since wildlife is such an important source of food and because it is not easy to substitute it in the forest zone with domestic animals, at least in the foreseeable future, any wildlife policy of West Africa governments that does not include wildlife utilization for food as its cardinal point is considered unrealistic. The problem of shortage of meat is more acute in West than East Africa, and a policy of wildlife development for tourism should not be allowed to take priority over its management for food.

CHAPTER 7

Wildlife utilization in East and Southern Africa

Introduction

For the purpose of students of wildlife conservation and protected area management, the following definitions are fundamental to the understanding of the concept of tourism and the contributions of wildlife-based tourism to national economies:

Tourism may be defined as the activities of persons traveling to and staying in places outside their usual environment, for not more than 1 consecutive year for leisure, business, and other purposes, not related to the exercise of an activity remunerated from within the place visited (CEC, OECD, VN and WTO, 2001). Tourism industries are hereby defined as groups of establishments producing the goods and services purchased by tourists (CEC, OECD, VN and WTO, 2001). Tourism industries in turn produce tourism products that can be characteristic (would not be produced without tourism), connected (partially characteristic), or nonspecific (general consumer products) (CEC, OECD, VN and WTO, 2001, p. 39).

Tourism consumption is the expenditure made by or on behalf of a visitor before, during, and after the trip and which expenditure is related to that trip and which trip is undertaken outside the usual environment of the visitor (CEC, OECD, VN and WTO, 2001, p. 39). Impact of tourism can be measured in terms of gross output, value added, and employment opportunities. In this context, the direct impacts (consumption) and the indirect impacts (linkages) and induced impacts (multipliers) are always considered in the assessment of economic significance of tourism. Economic activities involving the use of wildlife can be placed under two categories: consumptive and nonconsumptive uses. Consumptive uses include hunting, live animal sales, skins, hides, and other products. Nonconsumptive use includes all aspects of ecotourism, game viewing, photographic safaris, and other activities such as sportfishing.

In East and Southern Africa, nature-based tourism has grown tremendously over the last two decades. In countries such as Kenya, Botswana, Namibia, Tanzania, and South Africa, nature tourism is among the top

Wildlife Conservation in Africa
https://doi.org/10.1016/B978-0-12-816962-9.00007-7

earners of foreign exchange. In many sub-Saharan countries, the tourism industry is largely based on wildlife, where tourism is known to be one of the fastest-growing industries.

According to Mariki et al. (2011), tourism constitutes one of the fastest-growing sectors of the economy. Indeed, tourism is one of the top five categories of 83% of all developing countries (WTO, 1999). Consequently, it is an important industry in providing foreign exchange, employment, income, and promotion of rural and national development.

It is important to note that domestic tourism in Africa south of the Sahara is relatively low. Mariki et al. (2011) stated that in Tanzania, where wildlife- and nature-based tourism is well developed, domestic tourism in a well-developed Udzungwa valley in northern Tanzania was only 23% of the total tourist number. The low figure is typical of the pattern of tourist flow in most East and Southern African countries. Hence, African tourist trade relies mainly on foreign visitors.

However, in order to attract foreign tourists to a significant level, there are certain potentials, tourism standards in infrastructures, goods and services, etc. that African countries must be able to advertise to the outside world. Therefore, part of the potential of tourism development of African countries depends on their ability to attract foreign tourists. Furthermore, attraction of foreign tourists also depends on provision of the following well-developed basic infrastructures and ingredients for tourism (Table 13).

A comparative flow of international tourists to African countries

Table 14 presents a comparative international tourist arrivals and receipts in 49 African countries from 2010 to 2013. However, Table 14 does not show the national contribution of tourism to the gross domestic products of these countries.

The value of Namibia's wildlife assets

Namibia has one of the highest numbers of foreign tourist arrivals among the emerging economies in Africa (Table 14). In 2011, Namibia's tourist arrivals and receipts were estimated at 1,027,000 and 518 million USD, respectively. One of the factors responsible for a significant turnout of tourists was their spectacular tourism characteristic product, which is wildlife.

The physical wildlife asset (Table 15) is composed of an estimated 2.04 million larger wild animals, which produced a gross output of some USD 200 million and directly contributed USD 100 million to the gross national

Table 13 A checklist of some infrastructures that are basic to the development of international tourism.

(A) Specific products	(B) Nonspecific products
Tourism characteristic products	Goods and services
Accommodation	
Hotel and other lodging services	
Food and beverage services	
Restaurants	
Beverage-serving establishments	
Well-developed transport services	
Passenger transport	
Rail transport services	
Road transport services	
Air transport services	
Water transport services	
Support services	
Transport equipment	
Rail transport equipment	
Road transport equipment	
Rental	
Own vehicle	
Fuels	
Air transport equipment	
Travel agency, tour operator, and guide services	
Travel agency services	
Tour operator services	
Tourist information and tour guide services	
Cultural services	
Recreational and other entertainment services	
Tourism-connected products	
Goods	
Services	

product (GNP) (Barnes et al., 2009). Of interest is that the most significant component of wildlife use was nonconsumptive wildlife viewing tourism that generated some 62% of the total wildlife sector GNP contribution. Hunting tourism contributed some 19%, and live game production contributed some 10%. Other wildlife use activities contributing between 2% and 3% of the total sector GNP were meat production, intensive ostrich farming, and taxidermy. The wildlife use sector represented approximately 2.1% of national GNP in 2004 as compared with 4.6% for agriculture, 5% for fishing, 6.8% for mining, and 3.4% for tourism. In a further analysis growth in wildlife, use values over the next 30 years show the contributions from the sector could actually triple its economic contribution to the

Table 14 Relative flow of foreign tourists (international tourist arrival and receipts in African countries): available data on international tourism arrivals and receipts for African countries.

	International tourist arrivals (1000)					International tourist receipts (US$ million)			
	2010–2011	**2012**	**2013**	**2010**	**2011**	**2012–2013**			
1	Algeria	2070	2395	2634	2733	219	209	217	350
2	Angola	425	481	528	650	719	646	706	1234
3	Benin	199	209	220	231	149	188	170	
4	Botswana	2145				78	33	34	45
5	Burkina Faso	274	238		218	72	133	84	
6	Burundi	142				2	2	1	2
7	Cameroon	573	604	817	912	159	409	349	576
8	Cabo Verde	336	428	482	503	278	368	414	462
9	Central African Republic	54				11		11	
10	Chad	71	77	86	100				
11	Congo	194	218	256	297				
12	Côte d'Ivoire	252	270	289		201	141		
13	Democratic Republic of the Congo	81	186		191	11	11	7	1
14	Djibouti				63	18	19	21	22
15	Equatorial Guinea								
16	Eritrea	84	107						
17	Ethiopia	468	523	596	681	522	770	607	621
18	Gabon								
19	Gambia	91	106	157	171	74	83	88	

International tourist arrivals (1000)						International tourist receipts (US$ million)			
2010			2011	2012	2013	2010	2011	2012	2013
20	Ghana	931				620	694	914	853
21	Guinea				56	2	2	1	
22	Guinea-Bissau					13	9	7	
23	Kenya	1470	1785	1781	1433	800	926	935	881
24	Lesotho	414	397	422	320	25	29	46	39
25	Liberia					12	232		
26	Madagascar	196	225	256	196	321			
27	Malawi	746	767			33	34	34	
28	Mali	169	160	134	142	205	267	142	
29	Mauritania							48	41
30	Mauritius	935	965	965	993	1282	1484	1477	1321
31	Morocco	9288	9342	9375	10,046	6703	7281	6703	6854
32	Mozambique	1718	1902	2113	1886	197	231	250	241
33	Namibia	984	1027		1176	438	518	485	409
34	Niger	74	82		123	105	96	50	
35	Nigeria	1555	715		600	576	628	559	543
36	Rwanda	504	688	315	864	202	252	282	294
37	Sao Tome and Principe	8	12			11	16	13	13
38	Senegal	900	1001		1063	453	484	407	
39	Seychelles	175	194	208	230	274	291	310	344
40	Sierra Leone	39	52	60	81	26	44	42	59
41	South Africa	8074	8339	9188	9537	9070	9515	9994	9238
42	Sudan	495	536		591	94	185	772	773

Continued

Table 14 Relative flow of foreign tourists (international tourist arrival and receipts in African countries): available data on international tourism arrivals and receipts for African countries.—cont'd

		International tourist arrivals (1000)				International tourist receipts (US$ million)			
2010			2011	2012	2013	2010	2011	2012	2013
43	Swaziland	1078	879	1093	968	50	21	30	13
44	Togo	202	300	235	327	66	79	95	
45	Tunisia	6903	4785	5950	6269	2645	1914	2227	2190
46	Uganda	946	1151	1197	1206	784	959	1135	1184
47	United Republic of Tanzania	754	843	1043	1063	1255	1353	1713	1880
48	Zambia	815	920	859	915	125	146	155	224
49	Zimbabwe	2239	2423	1794	1833	634	664	749	851

Table 15 Physical wildlife asset count, 2004—estimated wildlife stock numbers in Namibia.

Species	Wildlife utilization zone 1	2	3	4	5	Total
Buffalo	1025	250	90	0	0	1365
Cheetah	706	149	405	270	2970	4500
Eland	1704	524	245	0	34,743	37,216
Elephant	9043	24	735	155	0	9957
Gemsbok	11,450	3115	18,670	5084	350,092	388,411
Giraffe	3683	229	666	68	5769	10,415
Hartebeest, red	1468	115	700	0	122,805	125,088
Hippopotamus	1262	0	300	0	0	1562
Impala, black-faced	1500	0	0	0	1870	3370
Impala, common	77	0	385	0	14,980	15,442
Kudu	2063	1484	1545	1000	345,801	351,893
Lechwe	0	0	250	0	284	534
Leopard	1970	430	960	640	4000	8000
Lion	574	23	109	22	0	728
Ostrich	3947	530	2840	2020	36,336	45,673
Rhino, black	816	43	45	75	134	1113
Rhino, white	54	62	0	0	75	191
Roan	440	120	95	0	435	1090
Sable	256	60	15	0	902	1233
Springbok	33,811	1771	37,150	37,270	621,561	731,563
Tsessebe	0	15	0	0	162	177
Warthog	148	61	40	0	173,866	174,115
Waterbuck	0	0	0	0	4475	4475
Wildebeest, blue	4985	224	470	0	16,623	22,292
Zebra, plains	18,098	0	20	0	7303	25,421
Zebra, mountain	8564	4347	2130	2175	55,520	72,736
Total	107,634	13,576	67,865	48,779	1,800,706	2,038,560

national economy and bring it close to its full spatial potential. Further increases in direct use value from wildlife will likely occur through intensive farming. Namibia's standing wildlife assets were estimated to have a value of NAD 10.5 billion (US 1.5 billion) in 2004, which is quite close to values estimated for fish and minerals (Table 15).

This is an Oryx a semidesert dweller common to the Kalahari desert of Namibia.

This is eland (*Taurotragus oryx*) the largest of all African antelopes. A mature male may stand at 6.5 ft shoulder with horns 3 ft long.

Wildlife species that are major tourist attractions in Africa

Figs. 6–13 show some wildlife species that are major attractions in Africa.

Specific economic value of wildlife-based tourism in Southern Africa

Hunting tourism, recreational and subsistence hunting that belong to the consumptive category, is widely practiced across Southern and East Africa, with hunting tourism or "sport hunting" regarded as being highly profitable. Yet, it is often said to be contributing little to national economies.

Table 16 shows that approximately US$3.2 billion was generated from nature-based tourism by 10 of the 14 Southern African Development Community countries (SADC) in 2000/01. This figure comprehensively includes all aspects of consumptive and nonconsumptive categories of wildlife-based tourism. In comparison, revenue generated by hunting tourism in seven SADC countries in 2008 was approximately US$190 million.

Although hunting tourism does not support large volumes of tourists, it can be regarded as the highest valued land use for arid and semiarid savannas in Southern Africa, especially in areas of low wildlife densities and diversity. Lindsey et al. (2007) estimated that a minimum of 1,394,000 km^2

Fig. 6 African buffalo or cape buffalo (*Syncerus caffer*). *(Photo by Olufemi Owolabi.)*

Fig. 7 See the legend in facing page

Fig. 7 Crocodiles (*Crocodylus niloticus*). Crocodile is the most dangerous and ferocious water dweller of African wildlife. Crocodiles are rarely found in the deep sea like the Atlantic Ocean. Some scientific research had been carried out on its reproductive ecology in order to achieve success in the attempts to reproduce them in captivity at the University of Ibadan as early as 1983 in the Department of Wildlife and Fisheries Management. Success had been achieved on the Nile crocodile *Crocodylus niloticus* under the supervision of Prof. S. S. Ajayi. The research include mating of Nile crocodile, collection of eggs, artificial incubation of the eggs, hatching of the eggs and raising the hatchlings to weaning age all in captivity. *(Photo by Olufemi Owolabi.)*

is used for hunting tourism in sub-Saharan Africa, which exceeds the area encompassed by national parks.

Table 17 summarizes the approximate gross value of hunting tourism for seven SADC countries between 2000 and 2008, with South Africa and Tanzania dominating the industry over this period. These data suggest that hunting tourism generates gross revenues of at least US$190

Fig. 8 Lion (*Panthera leo*). Lion is commonly referred to as the King of African jungle. Top: Is the group of lions lying, resting under the trees after they finish eating their kill. Bottom: The same group at the water-hole drinking after feeding. *(Photo by Olufemi Owolabi.)*

million per year (Lindsey et al., 2007) provided an estimate of US$201 million from a minimum of 18,500 clients).

Hunting tourism, recreational, traditional and subsistence hunting

Hunting across the Southern and East African region can be broadly categorized into three main sectors: (1) hunting tourism (also known as trophy or sport hunting), (2) recreational or biltong/meat hunting, and (3) traditional and subsistence hunting.

"Hunting tourism" or "sport hunting," also known as "consumptive wildlife tourism," involves the killing of wild animals in natural areas as opposed to "nonconsumptive wildlife tourism" where wildlife is not

Fig. 9 Leopard (*Panthera pardus*). Leopards are very powerful killers and don't eat their prey on the ground but on tree branches more than two (2) m above the ground. The arrangement of the black dots on Leopard is called rosette, like if one put together the five humans fingers into black paint and use it to the body on a regular space on the body.

physically killed. Here, the term hunting tourism is used in place of the terms "foreign hunting," "sport hunting," and "trophy hunting" and is defined as "undertaking hunting activities for one or more specimens of a certain species by a foreign or local hunter, who is willing to pay a fee for the special experience of hunting and/or attaining the trophy in a sustainable and ethical way."

Hunting tourism tends to be undertaken by foreign hunters that hunt for reasons of sport for a predetermined number of specific wildlife species, generally with the objective of keeping some part of the animal as a trophy (usually the horns or tusks of mature male herbivores or skins and skulls of carnivores). This form of hunting takes place on all types of land regardless as to whether it is owned privately, communally, or by the state. The foreign hunter will generally seek the services of a commercial safari operator and negotiate the cost of hunting a package of trophies in a given time. All species are hunted, with hunting packages for "charismatic trophies" such as elephant, lion, leopard, and buffalo fetching high prices. Hunting

Fig. 10 Cheetah (*Acinonyx jubatus*). Cheetah is the fastest African wild animal on four legs and is a Savannah grassland dweller. It has resemblance to the Leopard skin's external appearance but its own black and cream color configuration is single black dots. Cheetah is not a climber but Leopard is a perfect tree climber. *(Photo by Olufemi Owolabi.)*

of this nature is conducted under the supervision of a professional hunter regardless of the type of land involved and is strictly regulated by wildlife authorities (Booth, 2002/2009).

Socio-economic significance of hunting and angling tourism in United States of America and Europe

Comparison with the situation in the United States of America

Table 18 shows that in total, sportsmen in the United States spend approximately US$76 billion/year on hunting and fishing activities that generate approximately 1.6 million jobs and US$60 billion in federal and state taxes.

Fig. 11 Giraffe (*Giraffa camelopardalis*). Giraffe is the tallest of all African ungulates and Savannah dweller. *(Photo by Olufemi Owolabi.)*

Similarly, hunting is an important socioeconomic activity in Europe, particularly in rural areas. Kenward and Sharp (2008) estimated that within the European Union alone, hunting may be worth €16 billion (~US$22 billion) annually. This does not capture the complexity of values that can be added, derived, or offset from hunting activities such as the following:

Direct economic benefits—payments by hunters toward hunting fees, equipment accommodation, etc.

Environmental benefits—these include aspects related to conservation activities, ecosystem retention, and management.

Social benefits—hunting is often a very important social and cultural activity in many rural areas that has value in binding communities and offering useful social integration tools.

Fig. 12 African elephant (*Loxodonta africana*). Elephants are native to Asia and Africa. They are second to humans in habitat destruction and environmental modification. In view of the fact that they need about 6.4% of their body weight of food per day. They are also the easiest to be sighted in game viewing in wildlife base ecotourism destinations. Because of their size, they hardly get scared by the presence of human tourists. They were commonly sighted by game viewers in the 1970s to mid-1980s in the Kainji Lake National Park, New-Bussa, Nigeria and at Yankari, but now they are no longer easily sighted by tourists especially at the Kainji Lake National Park. *(Photo by Olufemi Owolabi.)*

Future bequest benefits—hunting helps retain a link to rural areas and an understanding of ecosystem functioning that serves to retain cultural knowledge and understanding for future generations.

Opportunity costs—these are related mainly to the costs that would be incurred by governments in managing wildlife populations and paying for damages incurred by these populations.

In summary, the experience of Southern Africa in consumptive and non-consumptive utilization has overwhelmingly shown that wildlife has contributed significantly to their national economies. However, illegal hunting and illegal trade in wildlife and other socioeconomic factors remain a threat to this vital continental asset.

Fig. 13 Hippopotamus (*Hippopotamus amphibius*). Hippopotamus lives in water with fish all around it but it is an herbivorous animal coming out of water at night to feed on vegetation. It is one of the location specific source of wildlife base ecotourism in many parts of Africa. *(Photo by Olufemi Owolabi.)*

Table 16 Nature tourism arrivals and income of SADC countries 2000–2001.

Country	Nature tourism arrivals (OOOs) (domestic and international)	Income from nature tourism (million US$)
Angola	0.9	0.3
Botswana	472.9	131.3
Malawi	109.4	13
Mozambique	42	8.4
Namibia	360	247.6
South Africa	4634.5	2298.8
Swaziland	243.9	27
Tanzania	203.7	299.9
Zambia	459.2	72.8
Zimbabwe	1494.4	143.5
Total	8020.9	3242.6

No data available for Democratic Republic of Congo, Lesotho, Madagascar, or Mauritius.

Table 17 Approximate gross value of hunting tourism in Botswana, South Africa, Namibia, Zambia, Zimbabwe, and Tanzania (US$ millions).

Country	Date	Gross income	Source
Botswana	2000	$12.6 million	ULG (2001)
	2008	$40.0 million	Martin (2008)
South Africa	2003/2004	$68.3 million	Damm (2005)
Namibia	2004	$9.6 million	Erb (2001)
Zambia	2002	$3.6 million	Child (2002)
Mozambique	2008	$5.0 million	Booth, unpublished
Zimbabwe	2000	$18.5 million	Booth (2002/2009)
Tanzania	2001	$39.2 million	Baldus and Cauldwell (2004)
	2008	$56.3 million	Booth, this study

Table 18 Approximate gross value of hunting and angling by sportsmen in Texas, Montana, and Wyoming in the United States (US$ millions).

State	Spendings	Jobs created	Salaries and wages	Federal and state taxes	Multiplier effect
Texas	$6.6 billion	106,000	$236 million	$1.4 billion	$11.6 billion
Montana	$721 million	11,500	$270 million	$520 million	$1 billion
Wyoming	$676 million	9500	$236 million	$137 million	$916 million

Source: Congressional Sportsmen's Foundation (CSF) www.sportsmenslink.org.

PART 3

Problems and strategies of conservation of wildlife in Africa

CHAPTER 8

Problems of protected (conservation) area management in Africa

Most of the problems associated with protected area management in Africa can be divided broadly into the following categories:

Ecological problems arise from natural causes and disasters such as drought, flood, erosion, outbreak of diseases and epidemics among wild animals, and migration and emigration problems associated with migratory species. So far, only drought and erosion have had deleterious effects on the management of biodiversity and wildlife in protected areas in Africa.

Fragile ecosystems—The problem of the Sahel and Arid Zones of Africa

One of the biggest ecological problems of natural resources conservation in the arid and semiarid areas and the Sahel Zone of West Africa is its ecological instability. The extreme influence of weather on the soil and the fierceness of the main climatic factors such as wind, sun radiation, and rain (whose duration is usually short and intense) are buffered by the vegetation. The vegetation cover, in an undisturbed situation, can take several years to attain a climax status, and this is when it can support an optimum biodiversity. However, the effect of increasing human and livestock populations with concomitant pressure on land resources has altered considerably or removed vegetation covers, to the extent that it can no longer mitigate the harsh effects of wind and sun radiation and occasional but heavy rainfall. The net result is wind and water erosion. This is a process of desertification that when combined with heavy population of livestock and humans, are the main causes of biodiversity loss in the Sahel Zone of West Africa—a phenomenon that is noticeable beyond latitude 12°N (Ajayi et al., 1998).

Wildlife Conservation in Africa
https://doi.org/10.1016/B978-0-12-816962-9.00008-9

Poverty—The bane of conservation in Africa

One of the biggest sociocultural and political problems with protected area management is poverty among the vast majority of African rural populations. In most countries of Africa, about 70% of the population live in rural areas and can be classified as "poor" who live under 1 or 2 dollars per day. These are children, women, and marginalized groups such as nomadic pastoralists and peasant farmers. They are alienated by the mechanisms of formal administration, and are not effectively integrated into policy and political process of development. This category of Africans live in closest proximity to protected areas. In most cases, they bear the social costs of natural resource conservation. Natural resources, particularly, nontimber forest products (NTFPs), remain their main sources of livelihood. In particular, they see wildlife and fish as cheapest alternatives to domestic livestock.

Around many protected areas, trees provide fruits and nut seeds, vegetables, condiments, spices, ropes, chewing sticks, edible oil, drinks, and flavors. In the tropical rainforests and savanna woodland of West Africa, over 150 indigenous woody plants have been identified as yielding edible products for man and animals. Furthermore, for many people in rural areas, the forests play the role of a "pharmacy" from which they obtain plant parts for treating their ailments. Economic forest products such as cane/rattan are used for the production of baskets, chairs, etc. Raffia products are used for the production of mats and ropes; wrapping leaves for food, fish poison, chewing sticks, honey, and mushrooms are also obtained from the forests. However, the rural people see these as their God-given resources that they continue to exploit for their livelihood—as they are too poor to afford the alternatives.

Past efforts have failed in most parts of Africa to recognize this fundamental fact and have placed too much emphasis on protecting "management areas" at the expense of human welfare. Various policies and legislative provisions were made to buttress this approach to conservation. These unfortunately have resulted into an attitude of indiscriminate exploitation of "protected areas," which in Africa interprets that what belongs to the government belongs to nobody. This problem of attitude should therefore be recognized in any policy formulation for protected area management in Africa.

Given the importance of forest products to the rural people, there was a widespread belief among rural communities that protected area management and conservation of biodiversity by its nature is a program of injustice. Furthermore, the "deprivation syndrome" they believe is in total disregard to African culture, which in many tribes is intricately interwoven with biodiversity.

The factors of poverty, disregard and disengagement from government projects among the rural populations, and the deprivation syndrome that are implicit in the past efforts in the management of protected areas have led to wanton exploitation and decimation of biodiversity. Any strategic plans for the management of protected areas in Africa that do not recognize and address this fundamental issue of rural communities and integrate them into management systems are therefore likely to fail.

Poaching and illegal trade in wildlife

Poaching and illegal trade in wildlife and by-products have been problematic to wildlife conservation in Africa since the beginning of the 20th century. Poaching is characteristically exploitive, nonselective, causing decline in the resource base, and drastically reducing the monetary value of wildlife to national economies (Ajayi, 1979). There are about three levels of poaching and illegal trades in wildlife that have been identified particularly in West Africa (Ajayi et al., 1979).

The rural dwellers live in closest proximity to protected areas. They exploit wildlife and other natural resources for food and cash. This category of people who turn to poachers engages in such illegal activities in order to "make ends meet." They engage in illegal hunting because of starvation and persistent poverty. They regard natural resources as their God-given assets, which, regardless of conservation ethics, form the basis of their livelihood. This is usually the main target group of rural communities for involvement and active participation in pilot projects on integrated natural resource utilization and rural development.

There is also another category of poachers who are licensed hunters. They reside in cities rather than in rural areas. There is little wildlife left outside protected areas, and therefore they deliberately invade protected areas. Their motive for illegal hunting is for trophies and the provision of "other varieties of meat" in their menu. Their poaching activities are not propelled by economic forces, but by social and cosmetic reasons as they belong to the upper or middle class of the society.

There is also another category of poachers. They are organized migratory and highly mobile poachers who are contracted to hunt for bushmeat and trophies by elitist business organizations, purely for economic reasons. They are financed by rich companies or individuals in urban areas to poach in protected areas where wildlife concentration is highest. This is the main channel through which wildlife trophies and captive live animals and other natural resources are smuggled out of Africa rather than through international airports and seaports where they can be recorded and taxed. Therefore, revenue from wildlife that should accrue to the economies of the countries concerned is usually insignificant, compared with the potential realizable monetary value of protected areas. The effect of illegal activities of this category of poachers is usually devastating to the natural resource base in protected areas and thus remains a bane of conservation.

Mitigation measures that are problematic to conservation

There are solution "pills" to management problems that themselves create further problems for protected areas. Most of the problems in this category arose from poor formulation; archaic, incongruous, and repressive legal instruments; ill-conceived management principles; and misapplication of management tools by protected area managers.

Weak, outdated policies and legal instruments

There are still some African countries where there is complete absence of a comprehensive policy for the management of wildlife and other natural resources. In many countries, there is no policy for the holistic approach to the conservation of biodiversity. Furthermore, even though there may be genuine efforts to manage protected areas and utilize natural resources for human welfare, these efforts are fragmented and uncoordinated. Thus, there is no natural resources institutional framework by which protected areas can be effectively managed and coordinated. Furthermore, examinations of some legislation that provide the legal basis for the management of protected areas in some West African countries show that "preservation" of natural resources rather than "conservation" is still the objective of their management efforts. There is also the problem of weak legislations and poor law enforcement, which prescribe token fees for offenders. For example, the penalty for killing a mature elephant in some African countries today and at the third millennium is only 55 cents, that is, 0.55 USD. These situations with regard to protected area management create serious constraints for field managers to the extent that they are rendered ineffective in the performance of their duties.

Another problem is that several existing legislations in several African countries are inapplicable or unenforceable for conservation of biodiversity. There is inadequate overall legal regime governing biodiversity and natural resource conservation. No legal mechanisms exist for participatory planning at all levels of government, which will determine priorities among competing resource uses and distribute costs and benefits fairly among them. Furthermore, most legislations were based on the defunct Organization of African Unity (OAU) Wildlife Algiers Convention of 1968. In consonance with the provisions of the convention, legislations should categorize animals into three "schedules": Schedule I is for those species that are endangered and are therefore to be strictly protected. Animals that are to be partially protected are in Schedule II. Schedule III contains those species that are unprotected and that could be hunted under license. However, in practice, a staggering list of species in Schedules II and III are endangered and are on the verge of extinction and should be placed in Schedule I. The problem here is not only the necessity for legal reforms but also the dire need for comprehensive national surveys to determine the true status of biodiversity in protected areas and what is left particularly of the wildlife resource in Africa.

Inadequate government support for conservation

There are numerous African countries that, through the lack of awareness of the value of conservation among government policy makers, give little support for protected area management. This has arisen mainly through ignorance of the value and significance of biodiversity. In some African countries, there is little or no provision for financing conservation and distribution of the benefits of conservation of natural resources to rural people and for rural development. Economic instruments are also not used effectively for achieving conservation objectives partly because economic development is progressing in many African countries without giving sufficient consideration to conservation. Furthermore, existing policies, where they exist, do not address the fundamental problem of poverty and the need to involve rural communities in conservation.

Cross-sectoral issues

Many countries have no established mechanisms for coordination and monitoring of protected areas, management programs, and conservation activities under different sectors and tiers of government. Furthermore, research and capacity building and planning are limited to few institutions with little coordination of efforts. Research, training, and planning are usually not incorporated into the mainstream of government policies. Public awareness of conservation issues is also very limited particularly in many African countries.

Local community participation as a tool for sustainable resource use

Until recently, emphasis in protected area management was placed on the importance of natural resources particularly wildlife, to the state rather than their significance to individuals, and rural dwellers who reside in closest proximity to protected areas.

Post World War II, conservation policies on protected area management tended to serve vital linkages, and participation by rural communities in decision-making, conservation, and utilization of biodiversity was limited or totally lacking. The rural communities had no access to the money and other benefits accruing from natural resource conservation in protected areas. This single factor was basic to the loss of their sense of responsibility for the welfare of natural resources. Communities in the vicinity of protected areas suffered severely from underdevelopment, mainly because money accruing from conservation of natural resources did not "filter" down to the grass roots. Consequently, these remote area dwellers (RADs) or rural communities have poor social amenities and suffer severe unemployment problems. The communities' interaction with their natural resources was often termed "illegal" under some repressive legislations. Since there was no involvement of rural communities in conservation process, natural resources continued to suffer from severe utilization pressure and decimation mainly from rural communities. It is against this background that various pilot projects were established, particularly in Southern Africa, literally to return natural resources to their "owners" so that communities can have (a) the custody, (b) the management, and (c) the utilization and benefits of natural resources, as they used to from time immemorial (Ajayi, 1990a,b,c). Indeed, what the Southern African countries were saying through the various community pilot schemes is one of the earliest messages from the beginning of times: "let them (man) have dominion over the fowls of the air, and over the cattle, and over every creeping thing that creepeth upon the earth."

Objectives of local community participation schemes in Southern Africa

Using the experience and demonstration projects such as the Communal Area Management Programme for Indigenous Resources (CAMPFIRE) in Zimbabwe, the Luangwa Integrated Resource Development Project (LIRDP), the Administrative Management Design (ADMADE) in Zambia, and the various wildlife utilization schemes in Botswana, these countries have established how dependent native Africans in rural areas are on biodiversity. The pilot projects in these countries have also demonstrated clearly how biodiversity can be utilized with people's participation as a tool for rural development.

Furthermore, according to Ajayi (1990a,b), the objectives of the pilot schemes can be summarized as follows:

(a) To improve the standard of living of the rural people who live in the project areas, through the sustainable use of the whole range of natural resource—forestry, wildlife, water resources, and agriculture

(b) To cover the cost of administration and equipment including all infrastructures and superstructures

(c) Without prejudice to (a) and (b) earlier, to provide revenue and other benefits at the national level

In summary, the intention is to increase the productivity and revenue-earning potential of the natural resources of the project areas and to redistribute the benefits in favor of the rural populations adjacent to the conservation areas.

However, as laudable as these intentions might be, there have been practical problems and constraints that constituted bottlenecks to realizing and reconciling the objectives of conservation of biodiversity with the utilization of protected areas for rural development. Some of the most intriguing constraints in protected area management and in reconciling the objectives of rural development with sustainable utilization of biodiversity require urgent, priority attention in Africa. These constraints are discussed later.

Constraints that limit the effectiveness of protected area managers from reconciling the objectives of conservation of biodiversity with rural development

Observations in this sector are partly based on the author's personal experience in many parts of Africa in protected area management and also on his personal experience when he led a mobile training seminary for United Nations/Food and Agricultural Organization (UN/FAO) in Malawi, Zimbabwe, Zambia, and Botswana (Ajayi, 1990a,b,c). The training seminar, which comprised 17 participants from the Southern African Development Coordination Conference (SADCC) countries, was to study integrated natural resource utilization with local community participation in these countries: Communal Areas Management Projects for Indigenous Resources (CAMPFIRE) in Zimbabwe, LIRDP and Administrative Management Design (ADMADE) in Zambia, and the Game Harvesting Project at Kedia in Botswana. The major constraints reported later are those associated with protected area management with rural dwellers as their main beneficiaries.

Lack of harmonization of conservation activities and ineffective institutional framework

One fundamental constraint with projects on natural resource utilization with rural community participation was the absence of harmonized legislation and institutional framework. Most of the projects in protected area management existed as "pilot schemes," as experimentations with enabling legislations. Furthermore, the results and impacts of these projects on rural populations and natural resources varied according to the institutions upon which the implementations of such schemes were based. It is therefore important to appraise the various mechanisms that will enhance the most successful implementation of these schemes and provide ideal institution and legislative "modules" for protected area management and rural development at local, subregional, and regional levels for sustainable conservation of biodiversity and rural development.

Poor financing of conservation activities in protected areas

Recent analysis suggests that government expenditures in this sector are grossly inadequate. In competing for government funds, conservation projects are often accorded low priorities in budgetary exercises partly because of the apparent lack of information on their socioeconomic values and also because the results of conservation efforts do not mature as quickly as most agricultural programs that yield quick returns. Failure to rectify this constraint could result in little growth in protected area management.

Dearth of research inputs

One of the major constraints to policy formulation, project planning, and implementation in conservation of biodiversity is inadequate research input and the lack of database. There is acute shortage of basic information on the reproductive and regenerative capacities of natural resources and on the trade-offs between alternative management options in land use schemes.

Inadequate capacity building

There is low capacity to conserve and utilize natural resources sustainably in protected areas partly because of inadequacy of trained personnel. There is an acute shortage of indigenous manpower in the application of key concepts and tools of scientific research. Most officers concerned with protected area management and conservation of biodiversity in government ministries and parastatal and nongovernmental organizations are not adequately trained to perform their duties. Despite the expanding requirements in systematic research input and training that have been created by recent developments in Africa and the global attention on conservation of biodiversity, very few credible, functional institutions exist at subregional or regional level that can adequately cope with human resource needs for development and capacity building. The alternative is foreign training with the attendant enormous costs and inappropriate training exposures. It is thus imperative to have several institutions for various levels of manpower training that will be supportive of conservation of biodiversity and rural development.

Specific training requirements for rural development

Sustainable rural developments require that people's needs are met, put that the natural resource bases are not destroyed or depleted. However, this should be accompanied within the realm of sustainability of the resource base and the environment. The need to increase villagers' participation therefore implied that people will have to be trained at the level on specific matters. The sustainable development of such schemes will require the participation of a multidisciplinary team of specialists (agriculturists, veterinarians, economists, social anthropologist, ecologists, etc.) who will be provided with special orientation training in the implementation of management objectives of protected areas and rural development.

Africa is a culturally and ecologically diverse continent. It becomes necessary therefore to provide a relevant set of targets for specific areas given the social realities and ecological possibilities in each country. This includes the natural resources and biodiversity, with its utilization possibilities for revenue generation and poverty alleviation. There is also a need for training of trainers in capacity building. Many development workers, particularly nongovernmental organizations (NGOs), are dedicated and committed to the goals and objectives of natural resource conservation and rural development. However, they lack the skills necessary to be effective in such schemes. The training programs and tools that are currently available to them are usually too generalized for specific localities. In this regard, it is necessary to evolve training modules that will be subregionally localized to ensure relevance to protected area management objectives and rural development.

CHAPTER 9

Principles for the management of protected areas

Characteristics of "protected areas"

"Protected areas" are a collective term for several categories of legally constituted areas by the state for conservation of natural resources. Management of wildlife and their habitats is however the prime objectives in most protected areas. According to FAO (1967), the characteristics of some protected areas are as follows: (a) "Strict nature reserve" is an area set aside to permit the face interaction of natural ecological factors without any outside interference whatsoever, except by competent scientific authorities. (b) The term "national park" refers to an area that is set aside for the protection of sites, landscapes, or geologic formations for the benefits and the enjoyment of the general public. Inside national parks, hunting, killing, or capturing of animals and the collection or destruction of flora are prohibited. (c) The term "game reserve" also denotes an area set aside for the management and propagation of wildlife and their habitats but in which hunting, killing, or capturing of fauna are prohibited and where human settlements and other human activities are "controlled" or prohibited. (d) The term "special reserve" or "sanctuary" refers to areas that are meant for special protection of characteristic wild animals and bird communities or for protecting threatened species of plants or animals and their habitats that are essential for their survival. In sanctuaries, all other human activities are subordinate to the above prime objective.

However, given the various objectives and criteria of each category of protected area, it is important to note that the African Convention for the Conservation and Management of Wildlife (1967) also stated inter alia "the contracting governments shall take advantage of the multiple values of their fauna, the management of which shall be an integral part of economic and social development planning and of land use policy." This declaration is fundamental to the concept of conservation of biodiversity in protected areas and rural development. It is also a concept that at the end of the 20th century, is generating debates on how to reconcile the set of objectives of conservation with those of rural development.

Wildlife Conservation in Africa
https://doi.org/10.1016/B978-0-12-816962-9.00009-0

Principles of in situ and ex situ conservation of biodiversity

In situ conservation is the on-site conservation of genetic resources in natural populations of plants or animal species such as forest genetic resources, in natural populations of tree and animal species. The process of protecting an endangered plant or animal species in its natural habitat is commonly known as in situ conservation. On the other hand, ex situ conservation is the relocation of endangered or rare species from their natural habitats to protected areas equipped for their protection and preservation. It is an essential alternative strategy when in situ conservation is inadequate.

This chapter deals with the definitions and principles of conservation that students of wildlife management, forestry, biological sciences, and agriculture and government and nongovernment organizations must know and have as tools for sustainable development of protected areas in Africa.

Principle of holistic conservation of biodiversity

By definition, biodiversity is the totality of genes, species, and ecosystems. It includes all heritable variations or differences in characteristics that exist in all living things, individuals, and their species in various ecosystems. In the late 20th century, it has become evident that the earth's biological diversity was disappearing, and it was exposed to a variety of threats. Species were becoming extinct at a rate unknown in geologic history, while the climate appeared to be subject to new phenomena.

One of the distinguishing features of the Rio 1992 Convention on Biological Diversity was that for the first time, a global consensus was reached to conserve all aspects of biodiversity: ecosystems, species, and genes (UNEP, 1995). Earlier conventions such as Ramsar on Wetlands and the Convention on International Trade in Endangered Species (CITES) on species in trade, etc. covered just fractions or aspects of the entire spectrum of biodiversity. Some aspects of the policy considerations on the Rio 1992 Convention on Biological Diversity are "sustainable use of biological resources, fair and equitable sharing of the benefits of conservation, and transfer of technology and funding mechanisms for natural resource conservation." What this means is a holistic approach as a basic strategy for sustainable development of protected areas. It is also recognized that biodiversity has provided man with his livelihood from time immemorial and that utilization of these resources should be conducted rationally and efficiently to ensure sustainability of the resource base.

At the global level, IUCN (1998) was concerned with the reconciliation of the objectives of conservation of biodiversity and rural development. It stated that management of protected areas should depend on a scientific process by which we understand nature and the impact of humans on nature and the analytic process of conserving the entire biodiversity as a natural system. The subobjective is the evolution of a political process that will marry various utilization objectives and thereby generate policies that will resolve conflicts between them.

Ecological principles

In order to "marry" sustainable use of biodiversity as a primary objective of protected area management with resource use of rural development, it is important that we draw on Southern African experience and from the "sustainable use principles" of the Southern Africa Sustainable Use Specialist Group (a division of IUCN Species Survival Commission, 1996). Utilization of natural resources is most likely to be sustainable when the following three ecological principles are addressed:

Conservation of ecosystems should be accorded a higher priority than the stability of individual species populations.

All species populations can be utilized for the welfare of man. There is no arbitrary population threshold below which use should be prohibited, if such use will be beneficial to the conservation of the species and the ecosystem.

The ultimate criterion of sustainability is the persistence of the species. Species population can be regarded as being sustainably used if their genetic and demographic characteristics under exploitation do not affect population viability. Expressed in another way, provided a species population is not reduced to the level of a real threat of extinction, then, the resource use can be regarded as sustainable. In ecosystem characterized by large environmental fluctuations, this may be the only criterion of sustainability.

Management principles

The most suitable technical approach to sustainable use of species lies in adaptive management. Adaptive management involves the following:

1. A basic hypothesis about the dynamics and functions of the system to be managed.
2. A clear statement of management objectives.

3. A monitoring system to provide the information required to modify the management systems, of the objective, or to modify the hypothesis if necessary.
4. In consumptive use systems, offtakes should often be tailored to reduce biological risks. However, this should be a pure precautionary management tool, and misapplication should be avoided. In competitive land use situations, injudicious use of the precautionary principle may result in disadvantage to conservation and overexploitation of biodiversity.
5. Utilization of natural resources often improves the status of the sued population, and this is the basis for the argument of use as a conservation tool. The corollary is that nonuse is risky—as species populations may decline to the brink of extinction.

Tenural principle

"Tenure" is one of the most important factors that affects sustainability of resource use. The term "tenure" covers all aspects of ownership, proprietorship, and rights of access to natural resources. Rights may be vested in the state, the community, or individuals. However, open-access resource systems are rarely used sustainably. Sustainable use can be attained when the following considerations are adopted for conservation of biodiversity:

(a) Rights of access should be clearly defined and accepted.
(b) There must be the ability to enforce those rights.
(c) The unit of management and accountability should be small, efficient, and functional.

Economic principle

The economic principle for conservation of biodiversity is based on correct evaluation of all species, that is, all species must have a value. Previous attempts to destroy market species and species products are not necessarily a panacea for conservation. However, sustainable use of species may rest upon the following factors:

i. The benefits derived from use are greater than the costs of conserving the resource.
ii. The rate of return from sustainable harvesting at least equals the return from alternative options.
iii. Short-term economic principles may provide incentive but can mitigate against sustainable use.

iv. There should be well-regulated legal markets with strong linkages to legal products.

Principle of sustainable use and rural development

The principle of rural development includes increase in per capital incomes of rural dwellers, improvement in nutritional and health status, improved achievements in literacy level, availability and access to public amenities, access to renewable natural resources, a fair distribution of income from natural resources, and a guarantee of basic freedom and fundamental human rights of rural communities.

One pertinent question to be asked is that if about 70% of the population of sub-Saharan Africa reside in rural areas and have been classified as "poor," given their intrinsic rate of population increase, can our depleted natural resources continue to support rural development indefinitely? For the increasing population growth and with little biodiversity left, it will be unrealistic to think that all requirements for rural development can be met through the exploitation of natural resources alone. Therefore, according to IUCN (1998), any sustainable rural development must be predicated upon a situation whereby decrease over time in natural resources should not be tolerated. Furthermore, sustainable management of biodiversity implies that we must progressively disengage ourselves from heavy reliance on the exploitation of natural resources for rural development.

That is why if biodiversity is to be conserved and utilized sustainably, it is important that African governments be urged to produce realistic, massive, and aggressive programs of sustainable rural development that will not rely solely on the utilization of biodiversity; this should be a priority consideration because about 70% of African populations reside in rural areas and are poor. The entire solution to their poverty predicaments certainly cannot be derived solely from the utilization of biodiversity.

Multiple uses for sustainable management of wildlife

The real advantages of wildlife as an economic and sustainable form of land use are realized in a system that offers a full range of uses such as recreational, aesthetic, nutritional, and scientific uses. A management system where several uses can be exploited simultaneously (combining all the above uses or part of) enhances improvements and efficiency of wildlife management, in competitive land use options.

The key point here is that multiplicity of uses of wildlife permits diversification of economic bases. Multiple uses of wildlife are practiced in several Southern African countries (particularly in Zimbabwe and the Republic of South Africa).

In these countries, "game parks," "game ranches," and "game farms" are managed purely for wildlife or in combination with livestock and crop farming. However, wildlife is subjected to a wide variety of uses as stated above, whereby meat can be produced sustainably and mote animals can be sold for substantial amount of money as trophies (and still providing meat that a hunting client does not want), thus allowing for safari charges to be made. Indeed, the most lucrative wildlife uses at present in Southern Africa is safari hunting and tourism. The landholders can also establish game-viewing enterprises, and since this does not require harvesting animals, it can support a large number that do not require harvesting animals. It can also support a large number of clients. The land is thus used for three tiers of wildlife utilization: consumptive meat production, lightly consumptive usage for trophies, and nonconsumptive tourism.

In addition to multiple use of wildlife itself, multiple use of the rangeland to support wildlife and livestock on sustainable basis is also being practiced extensively in Southern Africa. At moderate stocking levels, there is little competition between wild herbivores and livestock so that wildlife utilization can be added to extensive livestock ranching at little opportunity cost. In fact, combined wildlife and livestock operations are the most common form of wildlife utilization in Southern Africa today. In such combined enterprises, the wildlife component has been found to be much more profitable than livestock so that continuous operation of livestock does not make economic sense.

The point to note is that multiplicity of wildlife uses on game ranches, farm, and parks has allowed a full realization of economic value of wildlife, to the extent that this industry is fast replacing livestock production in Southern Africa.

Principle of natural resources conservation with local community participation in Africa

In the 19th and for the first half of the 20th centuries, most parts of Africa were colonized by Europeans. Large areas containing wildlife and other natural resources were set aside for "conservation" under state proprietorship. Relatively high densities of wildlife areas in many parts of Africa thus became the property of the colonial government. Rural communities that were traditional owners of wildlife resources were forcibly moved out of their ancestral areas of land, now called conservation areas and became alienated from wildlife that they once owned (Ajayi, 1990a,b,c). In practically all the wildlife legislations of the colonial and postcolonial governments, ownership of wildlife was vested in the government. The game laws discriminated against the local communities to the extent that they could not utilize wildlife for their basic necessities of life. The alienation of Africans from their wildlife with these conservation strategies was backed by repressive legislations with no respect to chiefs and rural communities and their traditional and cultural values.

Paradoxically, it was the very management strategy that the erstwhile colonial masters of Africa adopted to protect renewable natural resources that constitute the bane of conservation today. Large conservation areas were set aside for "protection" and "management" solely by government, with total disregard to the authority, welfare, and customs of the rural communities who live in closest proximity to the natural resources. The policy of "government's ownership" and "injustice" to rural communities was backed up by repressive legislations that led to complete alienation of rural dwellers from their "God-given resources." African chiefs soon lost their authority for the resources, while the communities were denied the benefits of conservation. The result of this management policy that had "natural resources, plants, and animals" as the focal points of conservation, with no regard to human welfare, led to wanton destruction of Africa's resources. Several African species of plants and animals have been lost through this management strategy, while a large number were reduced to a rare level of abundance.

This approach at conserving African wildlife and other natural resources had grievous consequences, as it cut into the heart of long-held traditions and customs that were an integral part of African existence. It has dismembered the entire holistic philosophy underlying the structure of African life. Deprivation has thus evoked a strong sense of injustice among rural communities to the extent that they resorted to poaching. The colonial governments and postcolonial governments responded to poaching by adopting vigorous militaristic antipoaching campaigns. Because this system prevented the African rural dwellers from using the resources that by custom was theirs by right, many of them resorted to large-scale persistent hunting and fishing, in defiance of the law. That was the origin of the much publicized "poaching" that has depleted most of African wildlife and other natural resources. The origin was a management strategy that in itself became problematic to conservation because it did not address the basic customs, psychology, traditional rights, and the dependence of rural communities that live virtually with and on wildlife and other resources.

Total ban on utilization and commercialization of wildlife

There are two approaches to elephant and rhino conservation in Southern Africa today. The first is a total ban on exploitation of these two species, in order to suppress or eliminate trade in ivory and rhino horns. This was the method adopted in most East and Central African countries, and it has resulted in the loss of 56%–78% of elephant populations in this region mainly to poachers. Kenya's population of elephant has also fallen from 65,000 in 1981 to 18,000 in 1989. It was under the same principle of conservation by prohibition that the population of rhino on the African continent dwindled to a meager 3500. This conservation approach and the rather catastrophic results were said by Simmons and Krenter (1989) to be a characteristic of an economic theory that teaches that a government ban on a valuable commodity can never wholly eliminate the demand. Hence, the total ban accomplished three results: price of the commodity increased—ivory and rhino horn; people with comparative advantage at avoiding detection, particularly government officials, took over the formerly legal market; and lastly, in the case of resources owned in common, the resources disappeared.

It must be emphasized that comparatively, so little was spent on patrolling the protected areas because the chances of catching poachers were minimal. Game guards were ill-equipped, underpaid, and sparsely scattered in reservation areas. Despite the total ban on exploitation of wildlife, the more lucrative business of poaching with such a powerful incentive to the very government officials that were charged with the primary responsibility of protection became the worst criminals in conservation history. After several years of this conservation strategy, East Africa lost about 56% of its elephants inside the park and 78% outside protected areas and by projection, under the so-called total ban approach. It is estimated that East African elephants may become extinct within a few decades. The failure of this management approach was attributed to (a) misplacement of responsibility of wildlife protection, management utilization to "government hands" instead of the local communities that reside in close proximity to the protected areas and (b) the inability of this conservation strategy to address the basic problem of starvation, poverty, and unemployment that are the root causes of illegal hunting among the local communities in Africa. The other approach, a positive one, is discussed later.

Quasi-private ownership and commercialization of wildlife

A report issued in June 1989 by the Ivory Trade Review Group (ITRG) concluded that it is the ivory trade and not habitat loss or human population increase that is responsible for the decline in numbers of African elephants. Some East African countries—notably Tanzania and Kenya—responded by an immediate total ban on the ivory trade. The European community correspondingly banned importation of ivory.

The East African countries also requested the secretariat of Convention on International Trade in Endangered Species (CITES) of wild flora and fauna to protect the elephant—with a total ban on trade in elephant products—including hides and ivory. There is now a shift from the old approach of total ban to that of conservation by utilization. This new approach, practiced by Zimbabwe, Malawi, Botswana, Zambia, and South Africa, allows ecologically based culling of elephants at the rate of 5% per year based on their intrinsic rate of increase, on private, communal, and state lands. Photographic and safari opportunities are also sold to international clients. This strategy has resulted into extensive marketing of elephants.

Under the utilization through quasi ownership and local community participatory strategies, the elephant population is now increasing in Zimbabwe, Malawi, Zambia, Botswana, Namibia, and South Africa—the countries that decided not to participate in total ban.

Also, contrary to the poor records of trade ban, wildlife commercialization has served as a protection strategy for a large variety of wildlife species. Zimbabwe farmers, for example, now ranch wildlife on a large scale, and crocodile farming is a multimillion dollar business in that country. Under the prohibition conservation strategy, white rhino declined from 1500 to 200 animals in five countries by 1960 (Simmons and Krenter, 1989). By contrast, South Africa's rhino populations, which were commercialized on private farms, parks, and reserves, increased 10-folds during the same period.

It is clear from this account that the solution to saving the African elephant and rhino and their sustainable management does not lie in total ban of ivory and rhino horn for all countries. Under strict control, utilization of elephants and other species of economic importance through quasi ownership, commercialization, and popular participation by local communities has led to a more effective conservation of wildlife. However, the success of the Southern African countries in this regard may be partly attributed to the availability of technical expertise and facilities for wildlife utilization schemes.

CHAPTER 10
Categories of conservation areas

The internationally accepted categorization of protected areas is presented in this chapter according to the International Union for Conservation of Nature and Natural Resources—IUCN—World Conservation Union. This chapter is based on the report of its special committee on criteria and nomenclature on national parks and protected areas (I.U.C.N., U.N.E., W.W.F, 1978).

Category I: Scientific reserve/strict nature reserve

Introduction

The rapid alteration of a variety of natural environments has created a need for a category of management that will ensure areas free of human intervention and unnatural outside influences and available exclusively for scientific research and environmental monitoring. These specific natural areas are necessary to provide adequate locations for research where a complete understanding of natural processes can be attained and where full protection of ecosystems and their inherent species is maintained.

Management objectives

The objectives of a scientific reserve are to protect nature (communities and species) and maintain natural processes in an undisturbed state in order to have ecologically representative examples of the natural environment available for scientific study, environmental monitoring, education, and the maintenance of genetic resources in a dynamic and evolutionary state. Research activities need to be planned and undertaken carefully to minimize disturbance.

Criteria for selection and management

These areas possess some outstanding ecosystems, features, and/or species of flora and fauna of national scientific importance. These areas are generally

Wildlife Conservation in Africa
https://doi.org/10.1016/B978-0-12-816962-9.00010-7

closed to public access, recreation, and tourism. They often contain fragile ecosystems or life-forms and areas of important biological or geologic diversity or are of particular importance to the conservation of genetic resources. Size is determined by the area required to ensure the integrity of such area, to accomplish the scientific management objective and provide for its protection.

Natural processes are allowed to take place in the absence of any direct human interference. These processes may include natural acts that alter the ecological system or physiographic features at any given time such as naturally occurring fires, natural succession, insect or disease outbreaks, storms, and earthquakes but necessarily exclude man-made disturbances. The educational function of the site is to serve as a resource for studying and obtaining scientific knowledge. Land use control and ownership should in most cases be by central government. Exceptions may be made where adequate safeguards and controls relating to long-term protection are ensured and where the central government concurs.

Category II: National park

Introduction

African governments have for some time recognized the desirability of setting aside outstanding natural areas representative of the diversity of ecosystems of their countries so as to guarantee their protection and use for present and future generations. Rapid exploitation of the natural resources has demonstrated that unless governments take positive and decisive action to protect the most outstanding examples of the country's natural heritage, these resources may be lost. The continued trend of urbanization has increased the need to provide opportunities for outdoor recreation and tourism in natural settings. Furthermore, the need for people to understand more fully the natural environment is of particular concern in an age of rapidly diminishing natural resources. Outstanding representative areas of a nation can serve to contribute to this understanding.

Management objectives

The management objectives of a Category II area call for the protection of natural and scenic areas of national or international significance for scientific, educational, and recreational use. The area should perpetuate in a natural state, representative samples of physiographic regions, biotic communities and genetic resources, and species in danger of extinction to provide ecological stability and diversity.

Criteria for selection and management

National parks are relatively large land or water areas that contain representative samples of major natural regions and features or scenery of national or international significance where plant and animal species, geomorphological sites, and habitats are of special scientific, educational, and recreational interest. They contain one or several entire ecosystems that are not materially altered by human exploitation and occupation. The highest competent authority of the country has taken steps to prevent or eliminate as soon as possible exploitation or occupation in the area and to enforce effectively the respect of ecological, geomorphological, or esthetic features that have led to its establishment.

The resource is managed and developed so as to sustain recreation and education activities on a controlled basis. The area is managed in a natural or near-natural state. Visitors enter under special conditions for inspirational, educational, cultural, and recreational purposes. The protected status of the area is adequately maintained directly by the central government or through agreement with another agency.

Category III: Natural monuments/natural landmarks

Introduction

Many countries possess natural features of particular scientific and educational interest; however, in many cases, they receive no special national recognition. These areas, however, may be susceptible to damage or adverse exploitation. The features might include outstanding examples of a country's natural heritage such as spectacular waterfalls, caves, craters, volcanoes, unique species of flora and fauna, and sand dunes. They would be of such scenic, scientific, educational, and inspirational importance that they merit special designation and protection. There is a need to elevate their significance because of their uniqueness so as to afford these areas greater protection for both scientific and public enjoyment.

Management objectives

Management objectives are to protect and preserve nationally significant natural features because of their special interest or unique characteristics and, to the extent of being consistent with this, provide opportunities for interpretation, education, research, and public appreciation.

Criteria for selection and management

This category normally contains one or more several specific natural features of outstanding national significance such as a geologic formation, a unique natural site, and animal or plant species or habitat that, because of uniqueness or rarity, may be threatened and should be protected. The specific feature to be protected ideally has little or no evidence of man's activities. These features are not of the size nor do they contain a diversity of features or representative ecosystems that would justify their inclusion as a national park. These areas have particular potential for public education and appreciation. Size is not a significant factor. The area should only be large enough to protect the integrity of the site.

Although Category III areas may have recreational and touristic value, they should be managed in such a way that they remain relatively free of human disturbance. These areas may be owned and managed by either central or other government agencies or nonprofit trusts or corporations as long as there is assurance that they will be managed to protect their inherent features for the long term.

Category IV: Nature conservation reserve/managed nature reserve/wildlife sanctuary

Introduction

Although most of the other categories of management play important roles in protecting and providing habitat for flora and fauna, it is essential that selected areas be set aside where manipulative management techniques can be applied to guarantee the stability or survival of certain species of plants and animals to assure their survival in the future. Due to widespread destruction of habitat, there is a need for specific areas that will protect breeding populations, feeding and breeding grounds, and critical habitat for protection of rare and endangered floral or faunal species.

Management objectives

The purpose of these areas is to ensure the natural conditions necessary to protect nationally significant species, groups of species, biotic communities, or physical features of the environment where they require specific human manipulation for their perpetuation. Scientific research, environmental monitoring, and educational use are the primary activities associated with this category.

Criteria for selection and management

A Category IV area is desirable when protection of specific sites or habitats is essential to the continued existence or well-being of individual biotic species, resident or migratory fauna of national or global significance. Although a variety of (protected) areas fall within this category, each would have as its primary purpose, the protection of nature, and not the production of harvestable, renewable resources, although this may play a role in the management of a particular area. The size of the area or in certain instances seasons, in which special management is necessary, will be dependent upon the habitat requirement or specific characteristics of the species to be protected. These need not require vast areas but could be relatively small consisting of nesting areas, marshes, lakes, estuaries, forest, or grassland habitats.

The area may require habitat manipulation to provide optimum conditions for the species, vegetative community, or feature according to individual circumstances. For example, a particular grassland or heath community may be protected and perpetuated through a limited amount of livestock grazing. A marsh for wintering waterfowl may require continual removal of excess reeds and supplementary planting of waterfowl food, whereas a reserve for an endangered animal may need protection against predators. These areas may be developed in limited areas for public education and appreciation of the work of wildlife management. Ownership may be by the central government or with adequate safeguards and controls in which long-term protection is ensured, by lower levels of government, nonprofit trusts or corporations, or private individuals or groups.

Category V: Protected landscape

In many areas of the world, distinctive landscape patterns are created by the integration of specific natural and cultural features that present esthetically attractive land and water settings. These may result through traditional land use practices of a cultural group in response to their environment, while in others, land use practices have retained relatively large and scenic natural or seminatural areas near urban centers.

At the same time, increasing population and leisure time and expanding urban areas are creating demand for additional recreation and tourism areas and facilities in esthetic environments for citizens and visitors.

The management and protection of these natural and cultural areas are important from both a recreational use standpoint and the scientific point of view because of their potential as reservoirs of genetic material and for their social customs evidenced through land use practices, which are disappearing under modern technology.

Management objectives

The objectives of this category are to maintain nationally significant natural landscapes that are characteristic of the harmonious interaction of man and land while providing opportunities for public enjoyment through recreation and tourism within the normal lifestyle and economic activity of these areas. These areas also provide for ecological diversity and scientific, cultural, and educational purposes.

Criteria for selection and management

The scope and character of areas that fall within this category are necessarily broad because of the wide variety of seminatural and cultural landscapes that occur within various nations. This may be reflected in two types of areas: those whose landscapes possess special esthetic qualities that are a result of the interaction of man and land and those that are primarily natural areas managed intensively by man for recreational and tourism uses.

In the first case, these landscapes may demonstrate certain cultural manifestations such as customs, beliefs, social organization, or material traits as reflected in land use patterns. These landscapes are characterized by either scenically attractive or esthetically unique patterns of human settlement. Traditional land use practices associated with agriculture, grazing, and fishing would be dominant. The size of the area would be large enough to ensure the integrity of the landscape pattern.

In the latter case, natural or scenic areas found along coastlines and lake shores or in hilly or mountainous terrain or along the shores of rivers and inland adjacent to important tourist highways or population centers offering scenic views and climatic variation are often included. Many will have the physical qualities and potential to be developed for a variety of outdoor recreational uses with national significance.

In some cases, the land would be privately held, and the use of either central or delegated planning control would likely be necessary to assist in the perpetuation of both the land use and lifestyle. Means of subsidization or other government assistance might be required for external renovations or construction to disguise improvements in the standard of living while recognizing the dynamics of evolution of the land and its use. Efforts would be made to maintain the quality of landscape through appropriate management practices. In other instances, the areas are established and managed under public ownership in perpetuity.

Category VI: Resource reserve—(interim conservation unit)

Introduction

Despite the rapidly increasing utilization of the natural resources in the world, there still remain land and water areas for which the most appropriate utilization has yet to be determined. If these lands are not protected, occupation and use are likely to occur on an unplanned, single-use, and short-term economic exploitation basis. This, in turn, may cause problems and compromises and limit future options where remaining resources need to be rationally utilized. This utilization without sufficient knowledge may result in resource deterioration and the loss of longer-term economic and social benefits.

Management objectives

One possibility is to restrict the use of these areas until adequate studies have been completed as how to best utilize these remaining resources. In such a case, the management objective is to protect the natural resources of the area for future use and prevent or contain development activities that could affect the resource pending the establishment of more defined objectives that are based upon appropriate knowledge and planning.

Criteria for selection and management

Category VI areas will normally comprise an extensive and relatively isolated and uninhabited area having difficult access or regions that are lightly populated yet may be under considerable pressure for colonization and greater utilization. In many cases, there has been little study or evaluation of these areas, and the consequence of converting these lands to agriculture, mineral or timber extraction, the construction of roads, etc. is unclear. Similarly, the use of the resources may not be appropriate because of the lack of technology, human, or financial resource restrictions or alternate national priorities.

Consequently, natural, social, and economic values are not sufficiently identified to permit the area to be managed for specific objectives or to justify its conversion to other land uses. Restricted access is implied so areas will normally require control, depending upon the pressures to enter and utilize the area. Lands may be government-owned, while others may be owned or administered by public corporations.

Maintenance of existing condition to allow for studies as to the potential use for the designated areas is a prerequisite. Protection studies and planning are envisioned as the major activities while under this short-term designation. No exploitation should occur with the exceptions of the use of resources by indigenous inhabitants. There is an acceptance of ongoing eco-sensitive activities.

Category VII: Natural biotic area/anthropological reserve

Introduction

In some countries, there may be a need for the protection of natural areas in which man is a component and obtains his livelihood by means that do not involve extensive cultivation or other major modifications of the vegetation and animal life. These individuals or societies may require special protection to maintain their existence.

Management objectives

The objective of management is to allow the way of life of societies living in harmony with the environment to continue undisturbed by modern technology. Research into the evolution of man and his interaction with the land would be a secondary objective.

Criteria for selection and management

Category VII areas are characterized by natural areas where the influence or technology of modern man has not significantly interfered with or been absorbed by the traditional ways of life of the inhabitants. These areas may be remote and isolated, and their inaccessibility may be maintained for a considerable period of time. The societies are considered relatively unique and may be of particular significance to the maintenance of genetic diversity and/or for research as to the evolution of man. These are predominantly natural areas of which man is an integral component. There is a strong dependence of man upon the natural environment for food, shelter, and other basic material to sustain life. Extensive cultivation or other major modifications of the vegetation and animal life are not permitted. Management is oriented toward the maintenance of habitat for traditional societies so as to provide for their continuance within their own cultural mores.

Category VIII: Multiple use management area/managed resource area

Introduction

There is a need for a category of management for areas that can both provide protection to natural resource systems and to certain natural resources and yet contribute significantly to a nation's economic, social, and material needs. The multiple functions of these lands can provide for a sustained yield of a series of products and services under proper management. Watershed protection, for example, may be of particular importance in addition to the timber, forage, or wildlife aspect of the area.

Management objectives

To provide for the sustained production of water, timber, wildlife, pasture, and outdoor recreation and at the same time provide for economic, social, and cultural needs over a long term. The conservation of nature is also an objective of this category primarily oriented to the support of the economic activities although specific zones may also be designated within these areas to achieve specific conservation objectives.

Criteria for selection and management

A large area contains considerable territory suitable for production of wood products, water, pasture, wildlife, and outdoor recreation. Parts of the area may be settled and may have been altered by man. Generally, these forests or other wildland areas do not possess nationally unique or exceptional natural features.

Planning to ensure the area is managed on a sustained yield basis would be a prerequisite. Land ownership would be under government control. Through proper zoning, significant areas could be given specific additional protection. For instance, the establishment of wilderness-type areas is consistent with the purpose of these areas as would be setting aside the nature reserves. Multiple use, in the context of Category VIII, is considered to be the management of all renewable surface resources, utilized in some combination to best meet the needs of the country. The major premise in the management of these lands is that they will be managed to maintain the overall productivity of the land and its resources in perpetuity.

Internationally recognized affi1iated designations biosphere reserve

Introduction

The UNESCO Man and the Biosphere Programs initiated 14 projects in 1970. The focus of project no. 8 is to conserve natural areas throughout the world through the establishment of a network of biosphere reserves.

Management objectives

The primary objective is to conserve for present and future use of the diversity and integrity of biotic communities of plants and animals within natural ecosystems and to safeguard the genetic diversity of species on which their continuing evolution depends. Biosphere reserves are to provide opportunities for ecological research, including, particularly, baseline studies, both within natural and altered environments. These reserves have particular value as benchmarks or standards for measurement of long-term changes in the biosphere as a whole and are consequently important sites for environmental monitoring. Biosphere reserves are to provide facilities for education and training.

Criteria for selection and management

Each biosphere reserve will include one or more of the following: (a) representative examples of natural biomes, (b) unique communities or areas with unusual natural features or exceptional interest, (c) examples of harmonious landscapes resulting from traditional patterns of land use, and (d) examples of modified or degraded ecosystems capable of being restored to more natural conditions. A biosphere reserve must have adequate long-term legal protection. Each biosphere reserve will be large enough to be an effective conservation unit and to accommodate different uses without conflict. Each reserve must be approved by the Man and the Biosphere International Coordinating Council before it can receive designation as a biosphere reserve.

Each biosphere reserve will be zoned to provide direction as to its management. Four zones may be delineated as follows: (a) natural or core zone, (b) manipulative or buffer zone, (c) reclamation or restoration zone, and (d) stable cultural zone.

World Heritage Site (natural)

Introduction

The International Convention concerning the Protection of the World Cultural and Natural Heritage (UNESCO, 1972) provides for the designation of areas of "outstanding universal value" as World Heritage Sites. These exceptional areas must be recommended by the signatory nation responsible for the site for declaration by the international World Heritage Committee. The sites will undoubtedly include many previously designated protected areas.

Management objectives

The primary objective is to protect the natural features for which the area was considered to be of World Heritage quality. Also important is to provide publications of information for worldwide public enlightenment and to provide for research and environmental monitoring.

Criteria for selection and management

Areas to be considered under the convention will be restricted to those that are truly of international significance. Natural sites must represent one or more of the following criteria:

(a) Be outstanding examples representing the major stages of the earth's evolutionary history
(b) Be outstanding examples representing significant ongoing geologic processes, biological evolution, and man's interaction with his natural environment
(c) Contain unique, rare, or superlative natural phenomena, formations or features, or areas of exceptional natural beauty
(d) Be habitats where populations of rare or endangered species of plants and animals still survive

Natural heritage sites must also fulfill conditions relative to the integrity of the site. Management of these sites will stress the maintenance of the heritage values; will ensure the continuation of legal protection; and will promote each site as to its significance to each country, its people, and the world. All sites will have to have strict legal protection and will be owned by government or a non-profit corporation or trust for the long term. While recreation and onsite interpretation will generally be developed, some sites may be of such significance that public use will either be strictly controlled or prohibited.

CHAPTER 11

Strategies for reconciling the primary objectives of conservation of biodiversity with rural development

Introduction

The resource regimes in sub-Saharan Africa are characterized by increasing human population, with heavy reliance on natural resources. This development is exacerbated by a myriad of problems and constraints of development that have been discussed earlier. These characteristics have led to declining biodiversity, declining productivity, and degrading quality of the environment.

According to IUCN—Sustainable Use Initiative (SUI) (1998), the term "sustainable use" symbolized a series of long-standing debates within the conservation community. There seems to be no clear-cut agreement on the precise definition and connotation of sustainable use of protected areas. However, after about five decades of protected area management in Africa, there are four pertinent questions that must be addressed with regard to sustainable development:

1. First, how do we reconcile the reality that people exploit natural resources and ecosystems for their welfare and livelihood, with our desire to conserve biodiversity in perpetuity?
2. What is the balance between the social and economic needs of the people and the needs for future generations and our genuine primary obligation to conserve biodiversity?
3. How far can utilization of natural resources cater for the needs of rural communities, and at what level of exploitation do we have a balance of nature and where protected area managers are obliged to discourage further intervention with nature?
4. In conservation, should we gradually disengage rural communities from their reliance on nature?

   In order to answer these questions, the foregoing management principles and recommendations have been proposed.

Wildlife Conservation in Africa
https://doi.org/10.1016/B978-0-12-816962-9.00011-9

Land use and management plan for conflict resolution

Management of protected areas is a scientific and administrative endeavor. When management of natural resources is based on a set of complex criteria and objectives, it is important to subject conservation to a dynamic process of effective planning. It is important that management of protected areas be based on a well-laid out management plan with adequate, reliable, baseline data on the biotic and abiotic resources that are available and the physical factors that influence the resources. The management plan will enhance sustainable uses of resources, remove conflicts, and ensure full realization and reconciliation of management objectives.

It is important to note that the spirit of Articles 8, 9, and 10 of the African Convention for conservation of wildlife (1967) (which has formed the basis for management of protected areas in Africa) is now outdated by recent global developments including the "Rio 1992 Convention on Biological Diversity." By this convention, ecological surveys and preparation of management plans should aim at conserving the totality of genes, species, and all ecosystems—that is, adopt a holistic approach to conservation of biodiversity. Future management plans should recognize that biological diversity has and will continue to provide man with his livelihood but only to a limited extent because of the rapid population growth in Africa. Therefore, rational utilization of resources for human welfare, rural development, and national development with local community participation should indeed form the cornerstones of future management plans. Furthermore, management plans, functioning as scientific tools and as first working "models" for protected area managers, should be based on the scientific principles of sustainable use.

"System plan" as a tool for reconciling management objectives

The word "system" connotes a "complex whole" of related objectives, interacting together to function as a unit. A "system plan" therefore refers to an administrative management design for coordination of a group of protected areas such as national parks, for realizing their management objectives. It ensures an organized approach to harnessing natural resource potential and also provides a systematic and sustainable conservation and utilization of biodiversity. A "system plan" places the complexities of characteristics of several protected areas under an institutional framework that will enhance multipurpose usage as consumptive and nonconsumptive uses of natural resources. Furthermore, the dynamic process of administrative and scientific designs in a "system plan" tends to reconcile seemingly conflicting objectives and ensures sustainable use of protected areas.

Trained manpower for the management of protected areas

Because of inadequate trained personnel, there is low capacity for the application of modern concepts of conservation of biodiversity and utilization of natural resources for rural development. It is necessary to have trained officers and personnel from village community level to managerial cadres who will be implementers of strategic plans for protected areas.

Policy reforms

There is a need particularly in West African subregion for evolution of all embracing, comprehensive policies for conservation of biodiversity, where protected area management will be prominent. The emerging policies should also embrace the following components:

1. A holistic approach to conservation of protected areas in which all species and ecosystems, whether they are of immediate use to man or not, should be adequately protected.
2. The international, national, and local significance of biodiversity.
3. The distribution of benefits of conservation to rural dwellers for rural and national development.
4. Introduce policies that will mobilize and involve local communities for active participation in conservation of biodiversity.
5. With population growth, higher level of literacy, and living standards, local patronage of conservation areas is likely to yield high revenue through non-consumptive use such as ecotourism.

International ecotourism has been the major revenue earner for several decades in some Eastern and Southern African states. However, there must be an appropriate national policy to uphold the ideals of Rio 92 Agenda 21 to preserve biological diversity and the principle of optimum sustainable yield in the utilization of natural resources and ecosystems. National policies must provide clear guidance on amount or percentage of total biodiversity to be utilized to prevent natural resources depletion. Finally, it must stipulate that most revenue and benefits are plowed back to the local communities in protected areas for rural development.

Legal reforms

There is a dire need for legal reforms in Africa—to update obsolete national/local legislations and also to enhance a holistic approach to conservation of biodiversity. The modernization of legal instruments should be based on current information on the status of plant and animal species, ecosystems, and important ecological formations. One of the prerequisites for legal reforms is a country-wide comprehensive survey to generate baseline data to determine the status of biodiversity.

CHAPTER 12

Local community participation in wildlife conservation in Southern Africa

Practically all the African countries that have been under some form of colonial influence went through experiences of alienation from wildlife resources conservation, deprivation from their God-given resources, and repressive legislation in their bid to conserve what is left of their wildlife resources. However, there is a growing awareness in Southern African countries today that wildlife and other natural resources cannot be managed successfully without the involvement and participation of the local communities (upon whose lands the natural resources exist), in the decision-making processes and in the distribution of the resulting economic benefits. Current programs in integrated sustainable wildlife management in Southern Africa therefore have, as their focal point, local community participation.

It was against this background that the communal integrated resource conservation programs were established in three Southern African countries: Botswana, Zambia, and Zimbabwe literally to return the natural resources to the "owners" so that communities have the custody, the management, and the utilization of their resources, as they used to do from time immemorial (Ajayi, 1990a,b,c). In other words, what Zambia, Zimbabwe, and Botswana have done in recent times through their community resource utilization schemes was in adherence to one of the earliest messages from the beginning of time: "Let them have dominion over the fish of the sea, and over the fowls of the air, and over the cattle, and over every creeping thing that creepeth upon the earth." This is the origin of the wildlife harvesting project in Botswana, the Luangwa Integrated Rural Development Project (LIRDP) and the Administrative Management Design (ADMADE) in Zambia, and Communal Area Management Programme for Indigenous Resources (CAMPFIRE) in Zimbabwe. This is the current philosophy on which sustainable management and utilization of wildlife resource revolves

Wildlife Conservation in Africa
https://doi.org/10.1016/B978-0-12-816962-9.00012-0

in Africa today. The current thinking of local community participation in wildlife management that has resolved many socio-political problems and that has reduced poaching to the barest minimum in many parts of Africa is therefore considered as the basis of modern wildlife development projects. The experiences of the three countries cited above in finding solution to their socio-political and legislative problem of conservation are presented below. The following was the account of the experience of the author of this book while he was a consultant to the United Nations/Food and Agriculture Organization (UN/FAO) in Malawi, Zambia, Zimbabwe, and Botswana in 1990.

The experience of Botswana game harvesting project

The objectives of the wildlife utilization projects were to establish game harvesting and craft production as part of the economic activities of the people who live in remote areas of Botswana and who are mostly farmers. The project was aimed at income generation and employment creation for rural communities. It was designed to be a low-input system, being guided by strict economic principles, requiring a minimum input of sophisticated technology so that it can break even within a very short time and allow the project to bear interest at a satisfactory rate. Revenue so generated as surplus or profit was plowed back to the communities for the development of that project and for other purposes decided by project participants.

The game harvesting project in Botswana can be described as a "marriage" between the communities and their wildlife. The community at Kedia was made up of Basurawa people who were legendary hunters and who have depended on wildlife as a means of livelihood from time immemorial. Therefore, the wildlife utilization scheme at Kedia was based purely on the premises of needs of the people, their hunting culture, and the welfare of the wildlife populations.

There were two wildlife utilization projects at Kedia: a game harvesting and a tannery. For the game harvesting project, the Department of Wildlife and National Parks determines the quota of each species of animals to be harvested each year by the community. The community was issued with a special license that is valid from April to December of each year during which they hunt through a few delegated hunters. Vehicles and guns were provided as a special assistance from the government of Botswana for the project. Marketable items that provided revenue from wildlife include meat, skins, and trophies. Revenue was shared in the following proportions:

70% was to the community (commercial bank account) and 30% to the project management account. The 30% revenue for management was spent on vehicle maintenance, ammunition, and staff salaries. Disbursement of the 70% of the revenue that accrued to the community was entirely at the discretion of the people themselves.

Benefits of the community projects to the people included employment opportunities, income generation, regular meat supply, skill development and improved wildlife conservation in terms of total elimination of poaching activities in this particular project area, increase in wildlife populations, and upgrading of their habitats.

In 1989, about US$20,000 accrued directly to the rural community representing 70% of total revenue for that year from game harvesting. For people who have not handled significant amounts of cash before, this amount of money to the Kedia community was phenomenal.

The experience of Zambia (LIRDP)

The objectives of LIRDP were as follows:

1. To improve the standard of living of the people living in the project area, through sustainable use of the whole range of natural resources in that locality, including agriculture, forestry, wildlife, and water resources.
2. The program recognized the principle of diversification of the economic base through the use of several resources to enhance sustainability and viability.
3. To cover the cost of administration of LIRDP projects and without prejudice to the above, to provide revenue and other benefits at the national level.

The intention was to increase the productivity and revenue-earning potential of the natural resources of the project area and to redistribute the benefits in favor of the rural populations adjacent to the conservation areas.

The merit of the project also lied in provision of employments to local communities, increasing their income and improving their standard of living, placing the responsibility of wildlife management and natural resources utilization in the hands of local communities, and thereby drastically reducing the impetus and incentive for poaching.

The administrative structure under which the LIRDP functions was a unique one. It was designed to coordinate government and nongovernmental projects as they relate to land and resource use and the full range of natural resources in the Luangwa Valley. The administrative framework

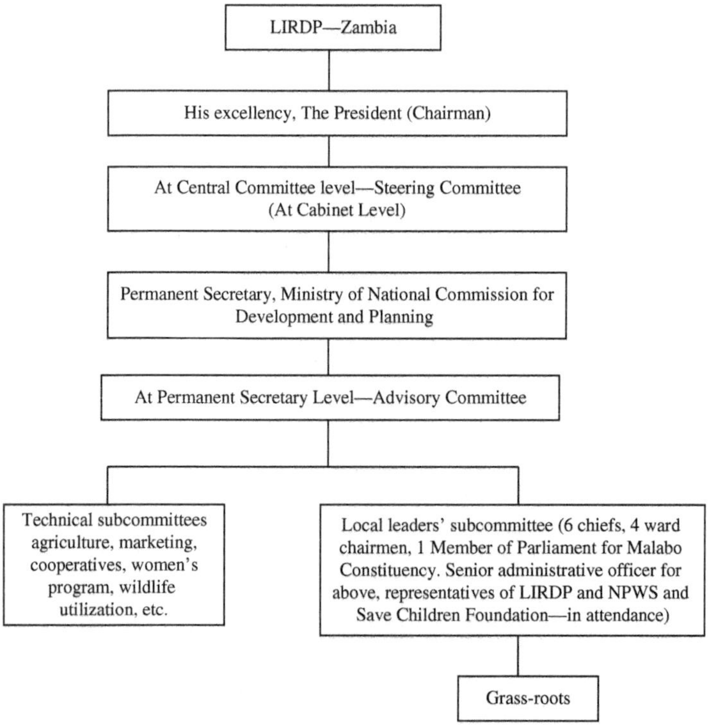

Fig. 14 Administrative framework of LIRDP in Zambia. The program aims at self-reliance and self-sustainable community-based resource management. *(From Food and Agriculture Organization of the United Nations. Ajayi, S.S. (1990). Training Programme on Integrated Wildlife Resources Use: rural community participation in Integrated Wildlife Management and Utilization in Botswana, Zambia and Zimbabwe. Commissioned by FAO: FAO/TCP/RAF/8962: Field document. Reproduced with permission.)*

also facilitates revenues from resource utilization to be reinvested through a revolving fund. The organizational structure (also presented on Fig. 14) is as follows:

The LIRDP is ultimately responsible to the president of Zambia.

An advisory committee operates under the permanent secretary of the National Commission for Development Planning (NCDP).

There is a local leader's subcommittee of the advisory committee, consisting of the six chiefs of the project area: the four ward chairmen, the member of Parliament for Malambo constituency, and the senior administrative office of Mambwe subdistrict. Representatives of LIRDP and the Department of National Parks and Wildlife Services (NPWS) and the Save the Children Federation are also in attendance.

Technical subcommittees of the LIRDP advisory committee exist in each of the technical areas within the mandate of the LIRDP programs. LIRDP management is also required to report to the provincial and district councils.

For day-to-day administration, the LIRDP coordinators are responsible to the permanent secretary of the NCDP.

Some salient features of the LIRDP projects

Under conventional wildlife conservation practices and before the creation of LIRDP, wildlife product values at about US$500m were extracted illegally from the Luangwa Valley in the past 30 years. One of the achievements of the present LIRDP system was the development of a viable and mutually beneficial relationship between the rural communities and their natural resources—through the above institutional/administrative framework.

The program aims at self-reliance and self-sustainable community-based resource management, thereby generating considerable revenue that is shared as follows: 40% for the communities and 60% for the running costs and management of projects. The community's six chiefs form the administrative base around whom the resource utilization and revenue-sharing revolves. The 40% allocation of revenue to the communities is given to the local leaders' subcommittee—and it is divided among the six chiefs. Disbursement of revenue is entirely dependent on the collective decision of the people through the local leaders' subcommittee. Tables 19 and 20 present a summary of revenues earned by LIRDP revolving funds in 1988 and 1989, which has been utilized for the following projects:

a. Construction of a clinic in the project area

b. Construction of a secondary school and staff houses

c. Purchase of a 64-seat AVM bus for local transport

d. Allocation of ZK105,000 to each of the six chieftainship accounts, uses of which were determined by the local communities

Table 19 Summary of revenues earned by LIRDP revolving fund 1988 (exchange rate: 10 ZK/US$).

	US$	ZK	60% ZK	40% ZK
Concession fees 1	30,500	305,000	183,000	122,000
Concession fees 2		7500	4500	3000
Animal license fees	70,000	700,000	420,000	280,000
1988 Total	100,500	1,012,500	607,500	405,000

Table 20 Summary of revenues earned by LIRDP revolving fund in 1989 (exchange rate: 16 ZK/US$).

	US$	ZK	60% ZK	40% ZK
Malambo Safaris	70,000	1,120,000	672,000	338,000
Concession fees	36,000	576,000	345,600	230,400
License fees	34,000	544,000	326,400	217,600
1989 Total	140,000	2,240,000	1,344,000	896,000

Anti-poaching

Before the project commenced, from 1975 to 1988, elephants were poached at the rate of about 10 animals per day, and thus, about 47,450 elephants were killed between 1975 and 1988. By 1990, there were only 5–10,000 elephants left within the Luangwa Valley, Game Management Area that has an area of about 14,000 km^2. There are also about 100 rhinos left in the entire region.

In 1989 when the LIRDP project started, the estimate of dead elephants was 32 that represented 0.01% of the annual rate of poaching from 1975 to 1988. Also, during the first year of operations of LIRDP, there were much fewer arrests of poachers, and less weapons were confiscated. This is an indication of a rapid decline in the incidence of poaching—a trend that continued till 1991.

This achievement was attributed to the involvement of the local communities in the decision-making processes with regard to integrated management and utilization of all the varieties of natural resources (including wildlife):

Total amount payable to the local leaders' accounts was 40%, that is, K: 1,301,000.

Plowing back of monetary benefits of conservation to local communities and withstanding (i) and (ii) above—the general improvement in the standard of living of the local communities through the integrated utilization of practically all the natural resources has contributed to minimizing the impetus for poaching.

The experience of Zambia (ADMADE)

It is believed that the emerging third world countries have shown a poor record for building conservation into their national planning priorities; it may well be because conservation as traditionally defined by the Western world does not address adequately the needs and aspirations of African rural communities.

ADMADE therefore does not emphasize conservation in the sense of restrictive legislation and law enforcement. ADMADE by contrast emphasizes the benefits and responsibilities for local residents traditionally prepared to share their lands with wildlife. ADMADE did not dictate the methods used to achieve reconciliation between wildlife and rural residents' interests. Instead, ADMADE provided an uncomplicated design to administer wildlife management through local participation and leadership, with guarantees that agreed revenues accrued from the resource return to local villages in the form of employment and community development. The way village participants used this design reflected their traditional respect for the resources, tribal customs, and appreciation of the legal options now available to them under the program. ADMADE allowed flexibility and adaptability in dealing with local issues concerning wildlife resources since solutions come from the local residents themselves.

The political and institutional framework

ADMADE had established for each Game Management Area (GMA) a local policy-making body consisting of the ruling chiefs, the locally elected political representatives of the government called ward chairman, the unit leader and other appropriate natural resource officers, and senior level government leaders, including the areas' member of Parliament, district governor, district political secretary, and the wildlife warden for that particular province. Also serving as members were managing directors of any company that has established a commercial interest for wildlife in the area. This body constitutes the Wildlife Game Management Authority or Area GMA.

When these authorities convened, customary values of traditional rulers were merged with technical views and the opinions of resource specialists. The chairman of this body is the district governor, the highest ranking government official in the region. Along with other government and party officials, his influence was enhanced in facilitating implementation of the recommendations of the authority.

A prerequisite to the establishment of wildlife management authority was that its Game Management Authority or Area GMA was able to provide sufficient income from wildlife annually to support such needs of wildlife management as its local village scouts program while also supporting projects identified by local residents for community self-improvement. The area from which these revenues were generated and utilized is designated an ADMADE Unit.

In most cases, units corresponded to GMAs. Exceptions existed where GMAs were too large to administer effectively as one unit. The authority's task was to adopt an annual program of wildlife management as recommended by its unit leader and to approve a budget for community development projects. These budgets must be within the agreed limits of the revenue-sharing formula adopted by ADMADE.

Sharing of revenue

The evolving fund revenues were allocated and disbursed the following year according to the following formula:

40% to the authority for meeting the unit's wildlife management costs
35% to the authority for financing local community development
15% to NPWS for supporting those ADMADE costs at the departmental level
10% to the Zambia National Tourist Board for the promotion of tourism

Distribution of income in ADMADE

It must be emphasized that the prerequisite to the establishment of a wildlife management authority was that its GMA must show the potentiality to generate sufficient income from wildlife annually to support such needs of wildlife management as its local village scouts program while also supporting the projects identified by local communities for community self-improvement. The areas from which these revenues were generated and utilized were designated as ADMADE units. Table 21 presents a summary of revenue that were actually collected under ADMADE, the sharing and utilization of such funds.

The experience of Zimbabwe (CAMPFIRE)

The origin of CAMPFIRE in Zimbabwe can be traced to the provision of the Zimbabwe National Parks and Wildlife Act of 1975 that stipulated that local authorities who were able to satisfy the Department of National Parks that they can manage the natural resources in their domain should be given the "power" or "appropriate authority" to do so. Benefits accruing from the management and utilization of these resources should not go into the government treasury, but should accrue directly to the district council and the rural communities (Ajayi, 1990a,b,c).

The District of Nyaminyami and, later on, Guruve were the first to be given such an authority by the government of Zimbabwe—to manage their own natural resources, which was mainly wildlife. Nyaminyami District Council

Table 21 ADMADE—distribution of income (US$) from hunting rights leases in the ADMADE program.

Year	1 Resource management and cost needs 40% US$	2 Local community development needs 35% US$	3 Resource management cost needs in adjacent national parks 15% US$	4 Promotion of tourism 10% US$	Total US$
1987	204,600	179,025.00	76,725.00	51,150.00	511,550.00
1988	184,995.20	161,870.80	69,373.20	46,248.00	462,488.00
1989	125,532.80	109,341.20	47,074.80	31,383.20	313,832.00
Total	515,128.00	459,737.00	193,173.00	128,782.00	1,287,820.00

Notes:
Retailed in the area it comes from—vehicle cost, village wildlife scouts, construction of camps, etc. Held in the management account of the wildlife management authority for the GMA.
Local chief—development account held by the wildlife management authority.
National Parks and Wildlife Service.
Zambia National Tourist Board.
Source: Ajayi, S.S., 1990. Training seminars on integrated wildlife resource use. *Arrangements for SADCC Training Seminars on Integrated wildlife Resource use*; Ajayi, S.S., 1990. Training seminars on integrated wildlife resource use. *Record of SADCC Mobile Training seminars on wildlife managements involving people's participation*; Ajayi, S.S., 1990. Training seminars on integrated wildlife resource use: rural community participation in integrated wildlife management and utilization in Botswana, Zambia, and Zimbabwe – *A collection of seminar papers*.

in 1985 demanded that they be given the appropriate authority and therefore their natural resources were relinquished to them the same year.

Institutional framework for CAMPFIRE

The administration of Nyaminyami District was through a district council known as Nyaminyami District Council. It consisted of elected councillors from each of the 13 wards in the entire district and the traditional chiefs (who were ex-officio members). The council had a primary responsibility to administer its own areas and to provide certain social services such as health and education for its people. Since Nyaminyami District Council got the responsibility to manage and utilize their principal resource that is wildlife, it went ahead to create a body known as "Nyaminyami Wildlife Management Trust" through which the management of the wildlife resource could be effected.

Nyaminyami Wildlife Management Trust had a constitution that made provision for membership to the grassroots level, that is, 13 ward councillors as representatives of the rural communities who have the custody of the wildlife resource and representatives of government ministries that were responsible for wildlife management and community development. It had an executive of professionals and administrative staff who administered the affairs of the Trust. The organizational structure is presented on Fig. 15.

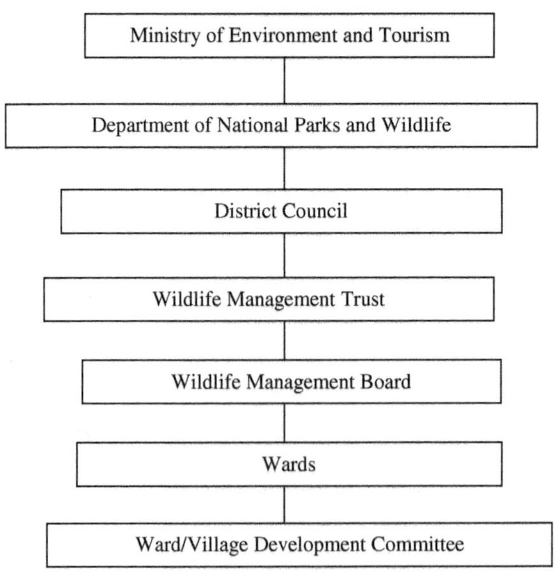

Fig. 15 Administrative framework for local community participation in CAMPFIRE projects in Zimbabwe.

The board membership and the involvement of the ward councillors ensured that vital decisions on wildlife resource utilization were taken by the rural communities themselves, while the government and a nongovernmental organization (NGO) Zim Trust maintained a supervisory role and provision of institutional facilities.

The Nyaminyami Wildlife Management Trust had a board of management responsible for the day-to-day running of the Trust. Membership for Kariba District is as follows: the project officer, Zimbabwe Trust, officials of the Department of National Parks and Wildlife, and some ward councillors. The chairman of the board of management is one of the ward councillors, who ensured again that decision-making was at the grassroots level, while government and "Zim Trust" officials maintained an advisory role.

Achievements of CAMPFIRE project at Nyaminyami

A prime benefit of CAMPFIRE was that the rural community had taken responsibility for conservation of wildlife. The government of Zimbabwe realized that wildlife cannot be conserved without the cooperation of the rural communities who live with these resources and bear most of the social cost of conservation. It would not have been possible for any community to respect a conservation project if their authority was not restored and their aspirations and security considered as integral parts of the package and terms of conservation. CAMPFIRE had ensured that the people themselves have the custody of the wildlife resource by policing and guarding it zealously at village level.

The result was that poaching is reduced to the barest minimum or eliminated completely in most places. Population of wildlife was rising at the same time that wildlife was being utilized by the communities. What this means in terms of conservation is that where the central government and its rifles have failed in "creating security" for wildlife, CAMPFIRE has succeeded.

There were also considerable revenue and other benefits: in 1989, about 400,000 Zim dollars (US$200,000) were realized from wildlife utilization, principally from safari hunting. Most of the money was spent to provide social amenities such as water, supply, provision of a village clinic, and a school for the Nyaminyami community. The chief of Molar, Chief Tombora, for example, was interviewed by the author in June 1990 on the benefits of CAMPFIRE to his domain and made it clear that the living conditions of his people have improved considerably with provisions of the rural infrastructures' physical and cash returns to his people that have been significant since CAMPFIRE programs took off 2 years before.

There was an "impala" cropping scheme whereby 500 "impala" antelope were cropped in 1989 alone and meat was supplied cheaply to the rural communities at about US$0.50 per kg, thus supplying animal protein to the people.

There was a crop damage compensation scheme whereby farmers who suffered crop damage by wildlife were compensated substantially. In 1989, about US$12,500 were paid as compensation to farmers within the area. In this way, farmers have built up their confidence in CAMPFIRE programs and reported crop damage to CAMPFIRE programs, since they were likely to be compensated for it. This is a conservation strategy that has reduced suspicion among the farmers and the management and has strengthened the community's support for CAMPFIRE projects.

Finally, it is noteworthy that officials of the Zimbabwe Trust maintained that the wildlife utilization scheme at Nyaminyami under CAMPFIRE was one of the most cost-effective projects in Zimbabwe.

Multiple uses for sustainable management of wildlife

The real advantages of wildlife as an economic and sustainable form of land use are realized in a system that offers a full range of uses such as recreational, aesthetic, nutritional, and scientific uses. A management system where several uses can be exploited simultaneously (combining all the above uses or part of) enhances improvements and efficiency of wildlife management, in competitive land use options.

The key point here is that multiple uses of wildlife permit diversification of economic base. Multiple uses of wildlife is practiced in several Southern African countries (particularly in Zimbabwe and the Republic of South Africa).

In these countries, "game parks," "game ranches," and "game farms" are managed purely for wildlife or in combination with livestock and crop farming. However, wildlife is subjected to a wide variety of uses as stated above, whereby meat can be produced sustainably and mote animals can be sold for substantial amount of money as trophies (and still providing meat that a hunting client does not want), thus allowing for safari charges to be made. Indeed, the most lucrative wildlife uses at present in Southern Africa is safari hunting and tourism. The landholders can also establish game-viewing enterprises. Since this does not require harvesting of animals, it can support a large number of clients. The land is thus used for three tiers of wildlife utilization: consumptive meat production, lightly consumptive usage for trophies, and nonconsumptive tourism.

In addition to multiple use of wildlife itself, multiple use of the rangeland to support wildlife and livestock on sustainable basis is also being practiced extensively in Southern Africa. At moderate stocking levels, there is little competition between wild herbivores and livestock so that wildlife utilization can be added to extensive livestock ranching at little opportunity cost. In fact, combined wildlife and livestock operations are the most common form of wildlife utilization in Southern Africa today. In such combined enterprises, the wildlife component has been found to be much more profitable than livestock that continuous operation of livestock does not make economic sense.

The point to note is that multiplicity of wildlife uses on game ranches, farm, and parks have allowed a full realization of economic value of wildlife, to the extent that this industry is fast replacing livestock production in Southern Africa.

CHAPTER 13

Suggested action plan for rehabilitation of wildlife in Africa: The case of Nigeria

The Chapter 2 of this book (i.e., on wildlife utilization in West Africa) shows the declining status of wildlife in West Africa. Hence, the rehabilitation plan presented in this chapter, though with Nigeria as a case study, is applicable to most West African countries.

Virtually, all Nigeria's wildlife heritage of about 274 species of mammals, 830 species of birds, and 90 species of reptiles is either at a rare level of abundance or on the brink of extinction. The rhinoceros, giraffe, ostrich, pygmy hippopotamus, gorilla, oryx, dama gazelle, dorcas gazelle, red-fronted gazelle, yellow-backed duiker, mountain reed buck, klipspringer, and Bates' pygmy antelope to mention a few, which, decades ago, roamed the forests and open savanna grasslands of Nigeria, are probably extinct forever. They remain mostly in the memories of elderly hunters and in African folklores. Hundreds of other species of wildlife are precariously endangered and are likely to vanish from Nigeria within the next few decades, unless an urgent step is taken to arrest and reverse the indiscriminate and wanton destruction of wildlife and their habitats.

Because of the precarious status of wildlife, it is important that West African countries embark on systematic survey and formulation of realistic action plans to rehabilitate their wildlife. Any program aimed at resuscitation of wildlife will translate to conservation of Africa's vital food and medicinal resources, its cultural heritage, and revenue base in ecotourism. In addition, the above species of wildlife, which have been the main sources of food and traditional medicines in Nigeria, particularly for those who reside in rural areas and who cannot afford the relatively high prices for meat from domesticated sources, are facing imminent danger of extinction.

Wildlife Conservation in Africa
https://doi.org/10.1016/B978-0-12-816962-9.00013-2

Specific objectives for rehabilitation of wildlife

The following are outlines of a proposal for a national survey and action plan for rehabilitation and restoration of Nigeria's wildlife heritage. The specific objectives of this proposal are as follows:

1. To salvage and restore, within the shortest period, the extinct, endangered, and largely decimated wildlife resources of Nigeria
2. To provide a well-articulated policy and legal instrument that will enhance effective and sustainable restoration of thousands of species of amphibians, reptiles, birds, and mammals that are imminently on the brink of extinction and the reintroduction of scores of wildlife species that have been totally exterminated in Nigeria
3. To engage the indigenous technical manpower in wildlife ecology in nationwide survey of all the ecological zones to determine the exact status of what is left of Nigeria's wildlife. To establish wildlife sanctuaries and rescue centers for endangered and endemic species (i.e., species that are only found in Nigeria and nowhere else in the world)
4. To relocate and multiply highly threatened wildlife species whose habitats have been drastically modified or destroyed or polluted through human activities such as agricultural and industrial developments and oil explorations
5. To produce an administrative design through local community participatory mechanisms for the sustainable reestablishment and rational use of Nigeria's wildlife heritage
6. To produce a robust, comprehensive action plan for in situ and ex situ development of wildlife in Nigeria and indeed in other parts of West Africa

Expected outputs from rehabilitation efforts

a. A plan of operation for the structures and systems, including technical, legal, and administration framework, will emerge within a short period of commissioning of this project to re-establish all the lost wildlife species and numerous endangered and highly threatened wildlife populations.
b. For ex situ conservation program, rescue centers, wildlife sanctuaries and zoological gardens, and specially protected areas will emerge by which effective and sustainable multiplication and re-introduction of species will take place.
c. An administrative design for active participation of local communities in all aspects of emergency rescue operations and rational exploitation of wildlife will also emerge.

d. In about 10 years of effective salvage operations, virtually, all the extinct and endangered species of wildlife will have been re-established in all the ecological zones of Nigeria.

One of the biggest causes of extinction of numerous wildlife species is intensive hunting. The main cause of over-hunting is poverty among rural Nigerians that live in closest proximity to wildlife habitats. This is because wildlife remains one of the cheapest sources of animal protein and it is seen as a cheap alternative to livestock and imported fish. This single factor has led to the disappearance of many wildlife species and the near-extinction status of several other species of wildlife. Therefore, any rehabilitation program that will re-establish and rejuvenate wildlife resources in West Africa must be based on a well-articulated policy that will involve the active participation of local communities. It must also address the basic problems of poverty.

Strategic plan for rehabilitation of wildlife

There are three main issues and legal problems that must be addressed aggressively for any rehabilitation exercise to be meaningful:

(a) There is a complete absence of coordinated national policy for wildlife rehabilitation and conservation in Nigeria. Despite the importance of wildlife in the national economy, the status of wildlife resources is still a mere appendage of Forestry Policy. It does not enjoy an autonomous status such as fisheries and livestock.

(b) There is also an absence of a National Wildlife Law to provide a legal basis for rehabilitation and conservation of what is left of Nigerian wildlife. What exists are several regional and state laws and edicts, some of which are as old as 1916. These regional legislations that were adopted by states that emerged from the respective regions are now archaic and do not address rehabilitation and modern conservation values.

(c) Furthermore, the penalties that were stipulated for offenders of wildlife exploitations are ridiculously low—sometimes prescribing a fine of equivalent of about US$30 cents for killing a mature elephant!

National inventory of wildlife

As a first step toward the arrest of wildlife extinction and to recoup what is left of wildlife resources, a 3-year period (three dry seasons and three rainy seasons) of a national survey of wildlife and their habitats is recommended for Nigeria. The nationwide survey will be mainly ground and aerial surveys using four-wheel drive vehicles and light aircraft and motorized boats where they are required. The following programs may evolve from the data and information collected from the national surveys.

Determination of the conservation status of each species of wildlife

i. Whether the species is totally extinct, endangered, or threatened in consonance with the International Union for Conservation of Nature (IUCN) criteria
ii. To determine the abundance and spatial geographic distribution of each species
iii. The economic, medicinal, and cultural uses of wildlife species, for example, for food, traditional medicines, and their cultural values and ecotourism
iv. The contribution of wildlife to national economy

A blueprint can then be developed within a few years for the conservation status, classification, and mapping of Nigeria's wildlife according to IUCN and natural resource criteria.

In situ conservation and creation of protected areas

Practically, all the "marginal lands" of Nigeria will be designated and mapped as conservation areas. Marginal lands are areas that are largely not easily accessible to man and where wildlife conservation as a primary form of land use does not necessarily compete with agriculture or any form of productive land use. A national survey can lead to save and rehabilitate Nigerian endangered wildlife species from extinction. The Nigerian Biodiversity Action Plan (1997) recommended that at least 20%–25% of its land mass be set aside for conservation of biodiversity, including wildlife. This approach will also be in consonance with United Nations Environmental Programme (UNEP) recommendations—to rehabilitate wildlife on at least 15%–20% of Nigeria's land mass instead of about 5%–8%, which is presently being managed by the National Park Service and the Forestry Department in Nigeria. Other programs that will emerge from the national survey and creation of protected areas all over the country are as follows:

(1) A program of regeneration of endangered species of wildlife in all the ecological zones of Nigeria
(2) Creation of special sanctuaries for reptiles, birds, and mammals in each of the ecological zones that will be accorded special protection for rehabilitation purposes under a new national wildlife law

(3) Rehabilitation of wetlands and lakes to provide sanctuaries to birds and waterfowls that are migratory and whose statuses are of international significance and whose habitats have been totally or partially destroyed

(4) Rehabilitation and re-establishment of important bird species of national and international significance such as Palearctic migrants, that is, birds that migrate from Europe to Africa in winter months and return to Europe during summer months, and for Afrotropical birds that migrate from East to West Africa and vice versa to avoid unfavorable food and climatic conditions

(5) Rescue operation for wildlife and habitat rehabilitation for animals that are currently facing imminent extermination as a result of habitat destruction through oil pollution particularly in the Niger Delta of Nigeria

(6) Rescue and rehabilitation of endemic species, that is, those species that are found only in Nigeria and nowhere else in the world such as mona monkey and drill in the Niger Delta and Rivers States, respectively

Ex situ conservation for rapid rehabilitation of wildlife through game ranching/rearing in captivity

A new approach for a rapid rehabilitation of wildlife is through game ranching. Nigerian scientists have perfected the rearing of the grass cutter (*Thryonomys swinderianus*), the African giant rat (*Cricetomys gambianus*), helmeted guineafowl (*Numida meleagris*), and the African giant snail (*Archachatina marginata*). Large-scale breeding centers can be established where the above animals, including crocodiles and small antelopes such as Maxwell's duiker and bushbuck, can be reared on a large scale as a source of food and revenue for rural and urban dwellers.

It is also suggested that production handbooks on the breeding techniques of these species be provided as a matter of urgency in major Nigerian languages for wildlife multiplication program as a means of restoring their populations in the wild and to promote large-scale participation and entrepreneurship among the unemployed Nigerians.

Establishment of large-scale wildlife ranches

In order to save the remaining wildlife populations from extinction, large game ranches under extensive management system can be established using antelopes such as duikers, waterbuck, kob, western hartebeest, roan antelope, and buffalo. Pilot projects can also be established in all the ecological zones (one pilot project in each ecological zone) as rescue and breeding centers, for wildlife rehabilitation and multiplication, and to facilitate animals' release to the wild after a successful multiplication program.

Development of zoological gardens and rescue centers

Zoological gardens can be created in major cities in the country (at least one for each ecological zone). They will serve multiple purposes as recreational and education centers and as rescue/multiplication ranches for representatives of Nigerian wildlife species, particularly those that are on the verge of extinction. Table 22 shows a summary of specific projects to be tackled with suggested time frame.

Establishment of zoological gardens will be based on master plans to be drawn by a team of zookeepers, ecologists, civil engineers, and architects. The master plan will incorporate the following components:

1. Site appraisal and land use planning for all the species of animals and the superstructures and infrastructures of the zoological garden
2. Architectural and engineering designs for animal enclosures and buildings
3. Techniques for animal acquisition and routine animal husbandry
   Rehabilitation and multiplication of animals for reintroduction back to their habitats on the wild

Table 22 Summary of specific projects to be tackled with suggested time frame.

Special action	Duration
A nationwide survey of what is left of Nigeria's wildlife heritage. Determination and classification of wildlife status according to IUCN criteria. Creation and mapping of protected areas to increase wildlife conservation areas from the present 5.8% to 25% of Nigeria's total land mass in consonance with UNEP guidelines for rescue operations for endangered wildlife species	3 Years
Establishment of six wildlife sanctuaries (one for each ecological zone) for rehabilitation of endangered and threatened wildlife species	3 Years
Establishment of six zoological gardens for multiple purposes of tourism, recreation, education, and multiplication of extinct and endemic wildlife species for reintroduction into their original habitats	3 Years
Rehabilitation of wetlands and unique wildlife habitats of national and international significance for migratory birds (Palearctic and Afrotropical migrants) and relocation of wildlife species that are facing imminent danger of extinction to suitable habitats. The Niger Delta, Rivers State, Mambilla Plateau, and Hadejia/Nguru wetlands are areas of international significance and sources of global concerns because of habitat loss to agricultural and industrial development and oil pollution	3 Years

Table 22 Summary of specific projects to be tackled with suggested time frame—cont'd

Special action	Duration
Production of national wildlife policy and national wildlife rehabilitation law, by a team of conservation lawyers, wildlife ecologists, and biodiversity experts	6 Months
Organization of two national stakeholders workshops for evolution of (a) a national policy for wildlife rehabilitation and (b) development of legal framework, a national wildlife legislation for effective restoration, and sustainable utilization of Nigeria's wildlife heritage	6 Months
6 Months	
Rehabilitation of Nigerian wetlands that are significant to the survival of wildlife, particularly in the Niger Delta and Hadejia/Nguru wetlands, which hold great promise for wildlife rehabilitation and also natural and artificial lakes and river basins for relocation of wildlife	2 Years
A vast program of wildlife farming of socially acceptable and highly prolific wildlife species with local community participation and in partnership with viable nongovernment organizations (NGOs), primarily for genetic conservation, utilization as a source of food, and reintroduction into the wild; species that have been completely domesticated and that can be multiplied are the grass cutter, helmeted guineafowl, bush fowl, African giant rat, African giant snail, and duikers. One wildlife farming center will be established for each of the ecological zones	3 Years
Establishment of two large-scale crocodile farms—for dwarf crocodiles in the Delta Region and for Nile crocodile in the Sudan Savanna	3 Years
Establishment of two large-scale duiker farms—one in the tropical rain forest ecosystem, for reintroduction and reestablishment of forest duiker, yellow-backed duiker, Ogilby duiker, and black duiker, and one in the Guinea Savanna zone for the establishment of revitalization of populations of Maxwell's duiker and red-flanked duiker	3 Years
Rescue/reintroduction and multiplication of near-extinct large antelopes, roan antelope, reed buck klipspringer, red-fronted gazelle, waterbuck, western hartebeest, buffalo, giraffe, and rhinoceros in the Guinea Savanna close to their original habitats to facilitate their return to the wild after a successful multiplication program	
Ostrich Multiplication Center: to be established in the Sudano-Sahel region of the country to multiply the ostrich and return them to their original habit	3 Years

CHAPTER 14
Migratory birds of Africa

Introduction

There are three major types of migration in birds:

Palearctic—Intercontinental migration from Europe to Africa and back every year

Afrotropical—Migration involving long distances across ecological moves within the same country and back and forth within Africa

Irregular—Migrating birds within the country

Palearctic migratory species

Palearctic migrants are intercontinental, migrating birds that fly in large numbers to escape the harsh winter months from the arctic region of Europe. They fly to the wetlands, lakes, and lagoons of West Africa and rivers where climatic conditions and food regimes are more favorable.

About 187 species of birds leave Europe and Asia on migration to Africa every year. Their arrival in Africa begins in August and reaches their peaks in November. Return journey begins in February and completes in April. Some Palearctic migrants spend most of their times in tropical Africa rather than in the temperate and semiarctic region of Europe and Asia. Of the 187 species migrating to Africa, 37 are the larger nonpasserines. Table 23 is a list of some Palearctic migrants usually found in Nigeria.

Afrotropical—African migration

Afrotropical or intra-African migration is a regular two-way movement of about 250 miles or more. Movement between breeding and nonbreeding range is performed annually according to seasonality. In West Africa where rainfall regime is a clear alteration of 3–4 months of dry season and 4–8 months of rainy season, depending on latitude, about 70 species of birds are known to perform regular north/south migration within West Africa and indeed within Nigeria.

https://doi.org/10.1016/B978-0-12-816962-9.00014-4

Table 23 Some Palearctic migrants.

Squacco heron	*Ardecia ralloices*
Marsh harrier	*Circus aeruginosus*
Western little sparrow hawk	*Accipiter erythropus*
Gray heron	*Ardea cinerea*
Western reef heron	*Egretta gularis*
Eurasian bittern	*Botaurus stellaris*
Northern shoveler	*Spatula clypeata*
Steppe eagle	*Aquila nipalensis*
Denham's bustard	*Neotis denhami*
Little ringed plover	*Charadrius dubius*
Arctic skua	*Stercorarius parasiticus*
Caspian tern	*Sterna caspia*
Royal tern	*Sterna paradisaea*
European swift	*Apus apus*
European bee-eater	*Merops apiaster*
Eurasian hoopoe	*Upupa epops*
European sand martin	*Riparia riparia*
Barn swallow	*Hirundo rustica*
Western yellow wagtail	*Motacilla flava*
Whinchat	*Saxicola rubetra*
European rock thrush	*Monticola slitarius*
European reed warbler	*Acrocephalus scirpaceus*
European pied flycatcher	*Ficedula hypoleuca*
European golden oriole	*Oriolus oriolus*

Some migrants, for example, the African swallow-tailed kite, *Chelictinia riocourii*, and the carmine and white-throated bee-eaters *Merops nubicoides* and *Merops albicollis*, breed in the northern parts of their range and migrate south to the moister parts of the dry season.The black kite, *Milvus migrans,* and red-necked buzzard, *Buteo auguralis*, migrate south in the dry season and breed in the southern part of their range and return to the north after breeding. Table 24 shows some Afrotropical (intra-Africa) migrating bird species.

Table 24 Some intra-Africa migrating species of birds that migrate from the south to the north of their range for breeding purposes.

African swallow-tailed kite	*Chelictinia riocourii*
Carmine bee-eater	*Merops nubicoides*
White-throated bee-eater	*Merops albicollis*
Black kite	*Milvus migrans*
Red-necked buzzard	*Buteo auguralis*
Standard-winged night jar	*Macrodipleryx longipennis*
Pennat-winged nightjar	*Caprimulgus vexillarius*

Irregular nomadic movement in birds

Most of the birds exhibiting nomadic movements are found in arid and semiarid localities in Nigeria. Their movements may not follow a regular pattern and are therefore unpredictable. They are described as nomads. An example of this is harlequin quail, *Coturnix delegorguei*, which breed in enormous numbers when local conditions are favorable. Examples are cinerea; chestnut weaver, *Ploceus rubiginosus;* and the fire-fronted bishop, *Euplectes diadematus*. The red-billed quelea, *Quelea quelea*, also perform irregular movements, breeding in enormous numbers. Some nomadic bird species are listed in Table 25.

Sensitive habitats of migratory birds in Nigeria

Out of the 13 open lakes and wetlands (Table 26), the highest concentration of birds are found in Nguru, Punjumu, and Dangona Lakes—which are the largest bodies of water in the wetland area. Dangona is already declared a waterfowl sanctuary.

As far as it is known, the only international convention for birds is "Ramsar Convention for the Protection of Wetlands and Waterfowls." Nigeria is not a signatory to this convention, and therefore Nigeria's wetlands and waterfowls are not bound by this convention. Nevertheless, it is important to note that about 70% of the 58,000 waterfowls of Hadejia–Nguru wetlands are Palearctic and Afrotropical migrants. They therefore deserve the highest international and national protection and surveillance.

Table 25 Species with irregular nomadic movements.

Harlequin quail	*Coturnix delegorguei*
Wattled starling	*Creatophora cinerea*
Chestnut weaver	*Ploceus rubiginosus*
Fire-fronted bishop	*Euplectes diadematus*
Red-billed quelea	*Quelea quelea*

Table 26 Sites in Hadejia/Nguru wetland area of Kano state that have been identified as the actual habitat and concentration points for migratory waterfowls.

Nguru Lake	Likori	Damasa
Daban	Kaban	Dumbari
Margadu	Matara Ganji	Punjum/Dangona
Kasago	Baturea	
Matafari	Kwikusha	

The possible propagation of H5N1 bird flu virus in Nigeria

It is important to note that the flight route of Palearctic migrating birds is through Western Europe—Finland, Sweden, Germany, France, Spain, and Portugal—and over Morocco, Algeria, and Libya and across the Sahara Desert to Niger, Burkina Faso, Mali, and Nigeria. It has also been established that wild migrating birds (particularly swans) have been found dead in Germany, France, and Spain—en route West Africa. The spread of bird flu H5N1 virus in these countries were immediately traced to migrating birds.

The first landing zones for migratory birds, particularly the waterfowls, across the Sahara are large bodies of water in countries bordering the Sahara such as Mali, Senegal, Burkina Faso, Niger, Nigeria, Chad, and Sudan.

In the case of Nigeria, the large bodies of water that they encounter after a marathon flight from Europe are listed in Table 27. The lakes and

Table 27 List of sensitive waterfowl habitants in Fadama sites in Bauchi, Kano, Katsina, and Sokoto States and the Hadejia–Nguru wetlands.

State	Local government	Fadama	Sensitive waterfowl habitats
Bauchi	Nagada	Bajoga	River Gongola and its plains
	Jamarea	Jamanee	River Hadejia and its flood plains
Kano	Kinikassama	Hadejia (Maza)	River Hadejia and its flood plains
	Tundun Wada	(Zangon Buhari)	River Tiga and Tiga Dam and Falgore Game Reserve
	Danbatta	Danbatta	Thomas Dam
Katsina			
Sokoto	Argungu	Yelema	River Argungu and its flood plains
	Talata	Brinin Tudu	The Bakalori dam, artificial ponds, and irrigation canals
	Mafara	Gidan Modi	Lugu Dam
	Wurno		
Hadejia–Nguru	Nguru–Hadejia	Nguru Lake Punjumu	Extensive Hadejia–Nguru wetlands: about 3365 km^2 waterfowl population is about 58,000; the most important waterfall habitats are Nguru Lake, Punjumu Lake, Dagoma waterfowl sanctuary, Daban Margadun Kassago, Matafari, Likoro, Kabau, Matari Gandu Daturia, Kurkume Damasa, and Dumbari

water bodies are in Kano, Yobe, Bauchi, Katsina, Sokoto, and Borno States. It is therefore not surprising that immediately after the H5N1 virus was detected in European countries on the routes of the Palearctic migrating birds, its presence was registered in Nigeria and later in Niger.

Fig. 16 shows some large bodies of water and wetland complex that provide the landing pads for migrating birds from Europe. These are the Hadejia/Nguru wetlands, which harbor about 58,000 birds, about 70% of which are migrating species. It is important to note that it was on a poultry farm in Kano State that bird flu was first detected in 2006 in Nigeria. There is every possibility that migratory Palearctic birds were

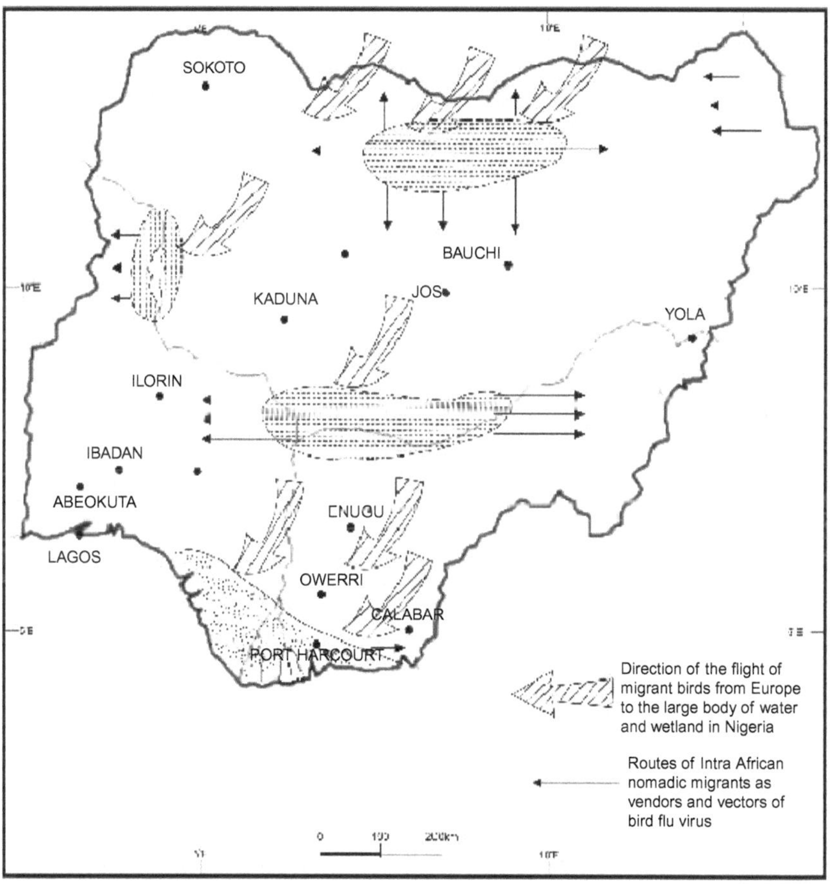

Fig. 16 Map of Nigeria showing major wetlands that are landing pads for migratory species. *(From Food and Agriculture Organization of the United Nations, Ajayi, S.S. (1998). Multipurpose Management of tropical rain Forest for bush meat production: A success story from West Africa. Paper commissioned by FAO – Forestry Development and authors contract Forest Management Division, FAO Rome, Italy. Reproduced with permission.)*

the main importers, while the intra-Africa and nomadic migrants are the vendors of the virus to poultry farms.

Of the 1000 waterfowls on the Oguta Lake in Imo State, about 60% are migrants from Europe. The Lake Chad, Yobe, Argungu, Kainji Lake, and the Niger Delta Creeks and wetlands are havens for migratory birds: osprey, gray heron, and pelican were once found dead on some islands in Rivers State and were subjected to laboratory examination for possible infection by H5N1 virus. This suggests the origin of the outbreak of H5N1 virus in Nigeria and as far South as Rivers State.

From the above account, there is a strong possibility that bird flu was introduced to Nigeria by migratory birds. By their nature, they do not observe international immigration and quarantine regulations when they arrive in Nigeria as possible importers of bird flu. Vendors such as Afrotropical or inter-Africa or intra-Nigeria migrants that are found in the same aquatic habitats as the Palearctic migrants can easily lend themselves as vendors and vectors of the virus to other parts of Nigeria.

Since the arrival of Palearctic migratory birds will continue till April (or May), that is, until the harsh winter of subarctic countries of Europe subsides, it is likely that March, April, and May are possible high risk months for bird flu in Nigeria and West Africa.

Furthermore, the role of intra-African and nomadic migrants as vendors and vectors of the disease should not be ignored. Unfortunately, the Federal Government of Nigeria has concentrated mostly on humans and domestic birds such as poultry for the prevention and spread of H5N1 bird flu virus.

Suggestions for the containment of H5N1 virus in Nigeria

It has been suggested that H5N1 virus was possibly brought to Eastern Europe from Asia where it first broke out. It has also been established that countries of South Western Europe such as France, Germany, and Spain were infected by migratory birds.

If there is any breakout of bird flu H5N1 virus in future, it will be expedient for the government of Nigeria and other parts of West Africa to set up surveillance team of wildlife ecologists to visit all the wetlands and bird sanctuaries throughout Nigeria that are known to harbor migratory birds. The team will provide first-hand information on the precise role of wild birds on the spread of bird flu and provide information on mortality of migratory birds and postmortem information on dead specimens.

Just as it is important not to be close to or be in contact with affected poultry birds, a public announcement should be made for people to avoid dead wild birds. The surveillance ecology will provide recommendations to governments on the protection and management of wild migratory birds in their habitats that will prevent the spread of bird flu.

The above accounts show the spread of bird flu by migratory wild birds in Europe, the occurrence of Palearctic migratory waterfowls in Nigerian wetlands, and the instance of bird flu first reported in areas adjoining the wetlands of Northern Nigeria. It is therefore important that the governments of West Africa shift their attention toward the surveillance of wild birds as vendors and carriers of H5N1 virus.

CHAPTER 15

In situ conservation of wildlife in West Africa

Introduction

In situ is a Latin phrase that translates literally to "on site" or "in position." It means "locally," "on site," "on the premises," or "in place" to describe an event where it takes place and is used in many different contexts. In situ conservation is the on-site conservation of genetic resources in natural populations of plant or animal species such as forest genetic resources, in natural populations of tree and animal species. The process of protecting an endangered plant or animal species in its natural habitat is commonly known as in situ. On the other hand, ex situ conservation is the relocation of endangered or rare species from their natural habitats to protected areas equipped for their protection and preservation. Ex situ conservation is an essential alternative strategy when in situ conservation is inadequate.

A multipurpose forest management with ethnomedicine, agroforestry, and bushmeat production—The Okomu Forest Reserve, Nigeria

Okomu Forest Reserve is situated in the lowland tropical rain forest of Edo State, southwestern Nigeria (Fig. 17). It contains some of the richest and relatively undisturbed high forest ecosystems in West Africa. The forest reserve is endowed with unique and rare biodiversity of global significance. It contains the relics of endangered wildlife species such as forest elephant *Loxodonta africana cyclotis*, chimpanzee *Pan troglodytes*, leopard *Panthera pardus*, and red-capped mangabey *Cercocebus torquatus* whose conservation status is of local and global concern. One species of forest monkey that is endemic to Okomu and is probably not found anywhere else in the world is the white-throated guenon monkey—*Cercopithecus erythrogaster*. In addition, there are about 33 other species of wildlife in the forest reserve that provide a significant portion of the animal protein consumed, particularly for the peripheral communities. Furthermore, traditional medicine, an important component

Wildlife Conservation in Africa
https://doi.org/10.1016/B978-0-12-816962-9.00015-6

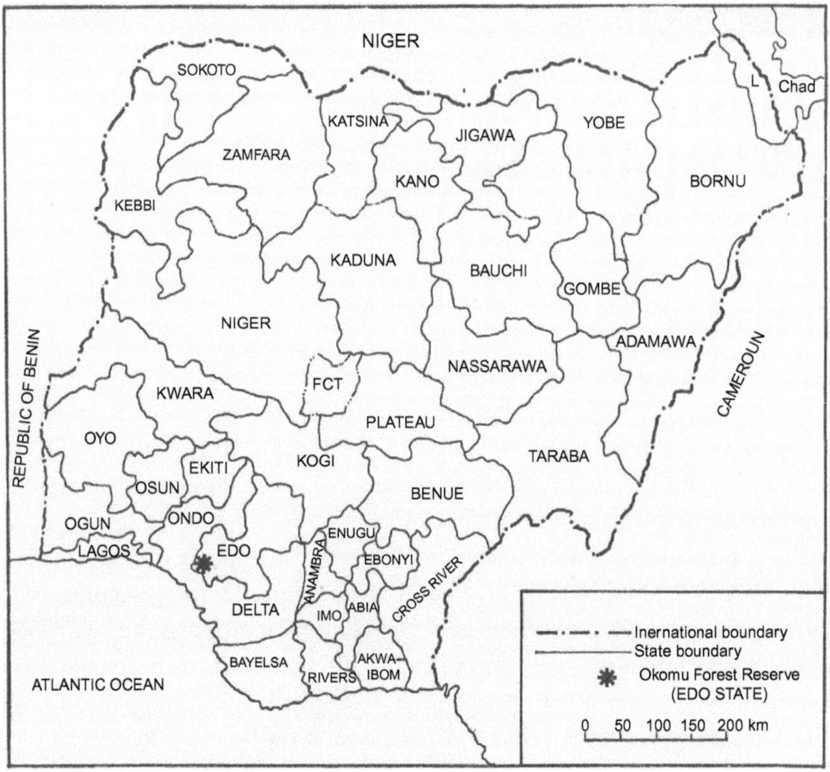

Fig. 17 Nigeria—showing the 36 states and location of Okomu Forest Reserve.

of the health-care delivery program in Southern Nigeria, is based on identified and confirmed medicinal plants in this high forest ecosystem.

However, in the past three decades, intense human activities such as industrialization, farming, logging, and hunting have led to a massive destruction of forest covers and the unique wildlife of Okomu Forest Reserve. Consequently, a foremost nongovernmental organization (NGO), the Nigerian Conservation Foundation (NCF) that is devoted to environmental protection, has since 1987 worked in collaboration with the Edo State Government for the sustainable development of Okomu Forest Reserve. The main objective of the various management strategies that are described in this chapter is to develop the protected area in partnership with the local communities, to create sustainable land use schemes, to include controlled hunting and logging, and to develop reforestation and farming schemes (including small game and fish farming), which are compatible with forest conservation.

The management is an example of a success story of a tripartite partnership effort between the government and an NGO and the local communities. This phenomenon is new to natural resource conservation efforts in West Africa, and it enhanced conflict resolution mechanisms and increased productivity of the ecosystem.

Location, climate and vegetation

Nigeria operates a federal system in which there are three tiers of government: federal, state, and local governments. National parks belong to the federal government, while all forest reserves and sanctuaries are the responsibilities of their respective state governments. There are 36 such state governments in which Edo State in southwestern Nigeria is one. Okomu Forest Reserve is located in Edo State of Nigeria about 75 km west of Benin City, its capital (Fig. 17). The protected area (forest reserve) has an area of 1082 km^2 between rivers Osse and Siluko to the east and west, respectively, and between longitude 5°E and 5°30′E and latitude 6°N and 6°N (Fig. 17). It is the largest and the least degraded forest reserve in Edo State. Most of Edo State, including Okomu Forest Reserve, lies within the lowland deciduous tropical rain forest. The reserve is representative of the rapidly disappearing lowland rain forest ecosystem of southwestern Nigeria and contains a high degree of biodiversity. It contains the endemic white-throated monkey (*Cercopithecus erythrogaster*), with other several animals that are on the Nigeria endangered species decree No. 11 of 1985, such as forest elephant (*Loxodonta africana cyclotis*), forest buffalo (*Synceros caffer*), yellow-backed duiker (*Cephalophus silvicultor*), and the African dwarf crocodile (*Osteolaemus tetraspis*). There is an impressive variety of bird life in the reserve.

Climate

The climate of Okomu is tropical, with well-marked rainy and dry reasons. The mean annual rainfall is 2100 mm per annum falling mainly between March and October; the driest months are December and January. The relative humidity is high, not less than 65% during the afternoons in any month of the year. The mean monthly temperature is 30.2°C. The highest rainfalls occur in June, July, and September. The dry season is preceded by two months of cold harmattan season in December and January, when the dry northeast winds from the Sahara desert prevail across the entire country.

Vegetation

The Okomu Forest Reserve is a semideciduous, humid, lowland forest. Freshwater swamp forests are found along the rivers. Mahoganies are well represented, and these include *Khaya ivorensis*, *Entandrophragma angolense*, *Ficus exasperata*, *Entandrophragma cylindricum*, *Guarea cedrata*, *Guarea thompsonii*, and *Lovoa trichilioides*. Other economic species are *Melicia excels*, *Gossweilerodendron balsamiferum*, *Terminalia ivorensis*, *Terminalia superba*, and *Triplochiton scleroxylon*. The forest is a mosaic of the following physiognomic forms:

"High forest" with well-stocked middle-land lower stories and numerous emergents

"Broken high forest" with small gaps interrupting the canopy at 6–12 m ("tall closed shrub") or 3–6 m ("low closed shrub") and a few emergent trees

"Open shrub" as above but interrupted by many glades containing light-demanding herbs

The extent of the scrub had greatly expanded due to gaps created by exploitation and by wind prone trees

History of land use

Okomu was legally constituted a forest reserve in 1912 by a Nigerian Government order 397. In 1950 it was reconstituted a Native Administration Forest under the Benin District Council. Before 1950, Okomu was all forests, relatively undisturbed. There were little farming activities, and the ecosystem represented one of the richest in biodiversity in West Africa. At this point, the reserve was divided into two areas BC 9 and BC 10 for management purposes.

In 1970 the reserve came under the management of the state government, to be managed on behalf of the local community. Despite the fact that the land was constituted a community land, to be managed by the state government on their behalf, the state government went ahead to grant a 99-year lease of 1500 ha from area BC 9 ha to an oil company and another 2000 ha from area BC 10 to a rubber company for development of oil palm and rubber tree plantations. Although each of these areas was termed as "dereservation," there was no evidence that the lease was ever legally gazetted. It is therefore not known how much this has affected the statutory rights of the occupancy.

In 1982 Oates and Anadu carried out a survey of conservation status of forest reserves in Ondo, Ogun, and Bendel States (part of Bendel State in

1982 is now Edo State). Because of the high rate of exploitations within the reserve and human settlements on its periphery, they recommended inter alia that a wildlife sanctuary be set up in the core of Okomu Forest Reserve to protect the unique flora and fauna of the tropical forest ecosystem (Figs. 18 and 19). In particular, it was thought that the creation of the sanctuary was the last hope for the protection of endemic and endangered species of wildlife. The creation of the sanctuary was gazetted in June 1988.

Management strategies

The rapid loss of the unique tropical forest biodiversity attracted the attention and participation of the NCF, which is a foremost NGO in Nigeria, in the management of Okomu Forest Reserve. Registered in 1982 as a charitable trust, the activities of NCF encompassed conservation education, support for wildlife research, and protection of endangered species.

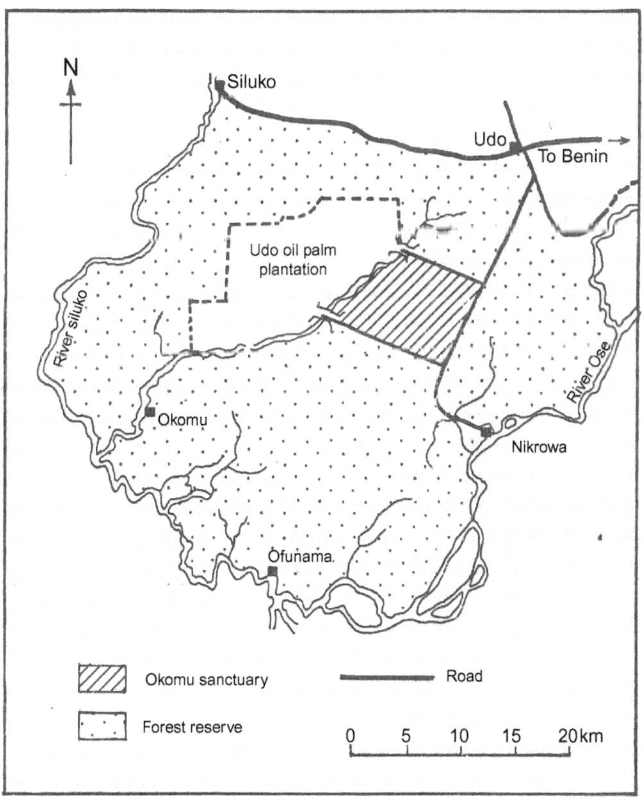

Fig. 18 Okomu Forest Reserve.

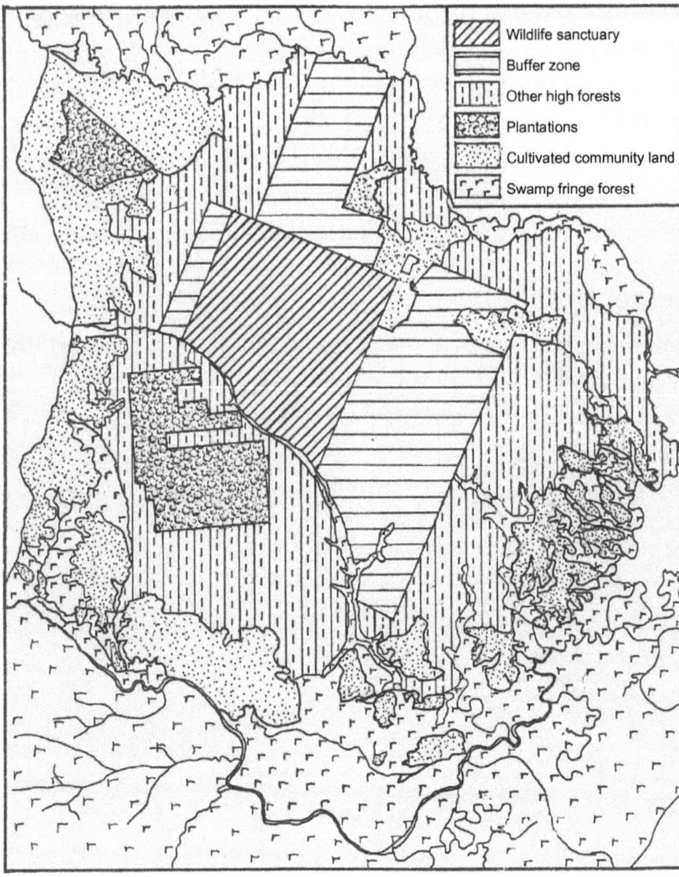

Fig. 19 Land use schemes in Okomu Forest Reserve. *(From Food and Agriculture Organization of the United Nations, Ajayi, S. S. 2011, Case study 2: Multipurpose forest management for bush meat production: a success story from West Africa, FAO, November, viewed 16 July 2015, http://www.fao.org/forestry/10258c60dbb6d55b4eb656bacabf3808aa4a3.pdf. Reproduced with permission.)*

The foundation is affiliated to the World Wide Fund for Nature (WWF), one of the largest private conservation organizations in the world. The main objectives of the management of Okomu Forest Reserve as stated in the memorandum of understanding between the Edo State Government and the NCF were as follows:

1. To protect the Okomu Wildlife Sanctuary as an inviolate refuge, for threatened plants and animals of the rain forest of southwestern Nigeria
2. To develop, in partnership with the government and the communities living in the project areas, sustainable land use schemes, including controlled hunting and logging; reafforestation; fish, small game, and crocodile farming; and agricultural practices that are compatible with forest conservation

3. To carry out an inventory and monitor on a long-term basis the resources of the sanctuary and surrounding areas
4. To generate support for wildlife conservation through promotion of conservation education among communities in the project area

Management inputs from Edo State Government

The Edo State Government was committed to the following obligations:

1. Provide guidance and support for the Okomu Forest Project
2. Nominate a local staff who shall assist the project manager in the execution of the project and shall be responsible for law enforcement in the project area
3. Appoint a conservation education officer who shall be attached to the project
4. Assign other cadre of staff as may be required for the effective protection of the project areas
5. To pay the full salaries and allowances of all officially assigned personnel
6. Ensure the fullest possible liaison between the project and the government
7. Implement, in collaboration with the NCF, rural development activities recommended by the project consultants

Participation by local communities

Management staff for law enforcement and labor within the reserve are recruited from the local communities.

The benefits of controlled hunting, agroforestry farming, and other benefits from the utilization of community farmlands and swamp fringe forests will accrue directly to the rural communities.

The program of land leases for farming activities was done with the participation of the local communities.

Multipurpose land use classifications: A management strategy for optimum forest production

The NCF carried out a complete ecological and socio-economic survey and produced a master plan for the management of Okomu Forest Reserve in 1995. The master plan contained six main land use zones as follows:

1. An inner conservation core—denoted a wildlife sanctuary
2. A forest/wildlife production buffer zone, contiguous to the sanctuary
3. Other high forests under long-term (99 years) community leases

4. A zone of plantations mostly for oil palm and rubber trees
5. Peripheral community farmlands with a practice of agroforestry.
6. The swamp forest fringes

Land use zonation as a management strategy

The above land use zonations served the following purposes:

First as a conflict resolution strategy among the major six land users in the forest reserve

For a total protection of the wildlife sanctuary that is the core breeding ground for wildlife for all the six land use schemes

Optimum utilization of the forest reserve by the rural communities and for commercial revenue generation to Edo State Government Forestry Department

Protection of biodiversity in perpetuity: using the strategy of multipurpose use with forest regeneration methods for sustaining forest integrity

Ensuring rational exploitation of wildlife for bushmeat and by-products and production and utilization of other nonwood forest products, such as fuel wood, leaves, and medicinal plants.

Details of land use practices and ecosystem management are presented on Table 28 and Figs. 18 and 19.

Socio-economic survey

Some officers of the Forestry Research Institute of Nigeria (FRIN) carried out a socio-economic survey of Okomu Forest Reserve in 1996 (Oguntala and Soladoye, 1996). FRIN has the mandate for forest research activities in all parts of Nigeria. The main objective of the socio-economic studies was to evaluate the importance of the forest reserve to the local resident peripheral communities and to Edo State Government. The studies included the structures of the local communities, the main nonwood forest products including bushmeat and medicinal plants and their monetary values, the food crops grown, and revenue generation estimates. The value of logging operations and land allocations to Edo State Government was also studied.

There were 14 village communities on the periphery of the reserve where a total of 90 questionnaires were administered. The total human population in the 14 villages was 1155. The sizes of the household sample ranged between 4 and 16. Data analysis was done using simple statistical tools, and the results were expressed as percentages and arithmetic means.

The results on Table 29 show that 56.28% of the village communities were fairly educated and 36.02% well educated, while about 4.9% were

Table 28 Multipurpose land use scheme of Okomu Forest Reserve.

Land use classification	Land use practice	Area (km 2)	% of all reserve
The wildlife sanctuary (an inner reservation core): denoted wildlife sanctuary	This block is for conservation of wildlife strictly for scientific and education purposes as the primary form of land use. Tourism is allowed on a limited scale, but there is total ban on hunting. The sanctuary was originally gazetted to protect an endemic white-throated guenon monkey (*Cercopithecus erythrogaster*), of which there was an estimated population of about 3000. Other primates present are the mona monkey (*Cehcopithecus mona*), the putty-nosed guenon (*Cehcopithecus nictitans*), and the white-necked mangabey (*Cercocebus torquatus*) whose populations were estimated at 170, 85, and 85 groups, respectively (each group size has 15–20 individuals). Galago (*Galago senegalensis*) and potto (*Perodicticus potto*) are very common, but exact population figures within the sanctuary are not known. There are healthy populations of buffalo (*Syncerus caffer*), and red river hog (*Potamochoerus porcus*). Populations of forest elephants (*Loxodonta africana cyclotis*) also exist in the sanctuary. Other species of animals include leopard (*Panthera pardus*) and chimpanzee (*Pan troglodytes*)	102	8.4
Forest production buffer zone: a zone of high forest around the sanctuary	The buffer zone is contiguous with the wildlife sanctuary. There is also a ban on hunting. Animals on the endangered species list such as elephant (*Loxodonta africana cyclotis*), chimpanzee (*Pan troglodytes*), white-throated guenon monkey (*Ceacopithecus nictitans*) are specially protected. Controlled timber extraction is allowed under license	235	17.5
Forest productions: other high forests under long-term community forest leases	This is an area where controlled hunting and logging activities are also allowed. However, as part of forest management practices, each logging contractor is required to plant five forest trees for every tree that is cut. In more peripheral forest compartment, adjacent village communities are given access to exploit the forest cover at all times. Area of lease per community is limited to 1 acre per head of total population of the area	180	13.4

Continued

Table 28 Multipurpose land use scheme of Okomu Forest Reserve.—cont'd

Land use classification	Land use practice	Area (km 2)	% of all reserve
A zone of plantations and other tree crop production (i.e., tree plantations and community forestry)	Community oil palm (*Elaeis guineensis*) and rubber tree plantations occur within the reserve. This is to boost smallholder tree crop production—to encourage smallholder development and provision of rural infrastructures. At the same time, the land use practice is geared toward forest regeneration and maintenance of rain forest mosaic. The land use planning encourages interactions between different zones. Therefore, zones are not mutually exclusive. For example, tree crops occur in community farmlands (described below), and farms occur within the plantations	202	15.1
Peripheral community farms with agroforestry practices	Human population community enclave lands occur mainly on the periphery of the forest reserve. Land use plan stipulates land allocation of 1 ha/person in each community, and therefore community land is commensurate to population size The present lease of 1 ha/person is based theoretically on subsistence agriculture per capital requirement under bush-fallow system. Communities that require more than allocation pay a certain "lease cost." Revenue that is so generated is plowed back to the community by the management for rural development. Communities are encouraged to keep small-scale livestock production based on goat keeping, snail, crocodile farming, and beekeeping for revenue generation for poverty alleviation Farmers practice agroforestry in consonance with forest regeneration and forest conservation policy. Where soil fertility is low, particularly in nitrogen, manure from livestock droppings is an option for intensive agricultural practice	164	12.2

	However, a more prominent, sustainable solution to soil infertility is by planting leguminous trees on community lands that will provide regular fodder for livestock, add nitrogen and organic matter to the soil, and the trees can be regularly coppiced to provide fuel wood Where the adjacent tree crop plantations have a good water supply, they act as regeneration agencies, using their nursery management expertise, to grow and sell tree seedlings to smallholder farmers for his agroforestry requirements For sustainable, multipurpose land use practice, the nurseries also cater commercially for the needs of the community forests and buffer zone tree regeneration		
The swamp forest fringes	The extensive swamp forests and river fringes are the main areas where raffia palms have been planted for the production of palm wine and local gin. The swamp forests are important for fish and wildlife diversity. Logging activities are therefore carefully controlled to avoid destruction of the environments for fish, wildlife, and humans. Management policy requires that swamp forests remain so by planting more raffia palms and swamp forest trees. Logging is scattered throughout the swamp forests There is an annual inspection of the swamp forests by an independent neutral agency. Penalties for not adhering to strict conservation and environmental protection guidelines include an upward review of lease costs to reduce profit margins of the village communities	297	22.2
Total	1340	100%	

Table 29 Literacy composition of the 14 village communities sampled.

Gender	Fairly educated	Uneducated	Totally unenlightened
Male	243 (21.04%)	46(3.9%)	21 (1.73%)
Female	177 (15.32%)	105 (9.09%)	30 (2.6%)
Children	230 (19.19%)	265(22.59%)	39(3.38%)
Total	650 (56.28%)	416 (36.02%)	89 (4.9%)

totally unenlightened. It is also interesting that more men were educated than women. The low literacy percentage is a reflection of the fact that forest and farm labor was the main preoccupation of these peripheral and enclave communities. Another study also confirmed that women provide a higher percentage of such farm/forest labor than men.

Results

The importance of most common items of food crops grown and consumed or sold for revenue generation by the village communities is also shown in Table 30.

The study showed two important features. First was that they grew most of the food they needed, and that was why they relied almost entirely on forest land for their survival. Secondly, they had limited food alternatives as those listed on Table 30 were almost exhaustive of their food choices. This was partly a result of the tropical rain forest environment and relatively poor soils that were more favorable to root crops than grains. Another explanation is their low literacy level that impaired their knowledge of food balance requirements. Most of the populations were farmers and forest workers who require high-energy foodstuffs such as the root crops they grew to support their daily activities.

Table 30 Main food items grown by the village communities.

Food items	% Respondents with "yes" affirmation	% Respondents with "no" affirmation
Yam	83.33	16.67
Cassava	86.67	13.33
Plantain	90.00	10.00
Maize	48.89	51.11
Cocoyam	46.67	53.33
Vegetables	28.89	71.11
Others (rice, beans, etc.)	57.78	42.22

According to Essang and Olajide (1974), land, capital, labor, entrepreneurship, and moisture are the five factors that mostly influence agricultural production in Nigeria. Therefore, one of the objectives of the study was to compare the relative importance of different land acquisition systems between 1978 and 1998, that is, a period of two decades. The results are presented on Table 31.

The results are expressed as aggregates for all sampled plots.

The following characteristics were observed on the land tenure and acquisition systems: By far, the most important system about 3 decades ago was government land allocation, and this was because the land was and is still a government forest reserve. However, community allocation particularly among the enclave communities has become a major rival to government allocation. Land acquisition through inheritance and rentals are however common among the peripheral communities. Land allocation through taungya system varied from community to community but ranged between ½ and 2.0 ha/head at a fee of 2–4 dollars/ha.

The four major products that the communities derive from the forests are fuel, wood, leaves, bushmeat, and wildlife by-products including the large forest snails (*Archachatina marginata*). The monthly quantities and their monetary values are presented on Table 32.

The annual revenue from leaves and fuelwood was $12,432.00 and from bushmeat, including snails, was $12,516.00.

Most of the residents were farmers. Their main crops and the monetary values are listed on Table 33. The units of evaluating the quantities were ropes for yam, basins for gari (processed cassava flour), lorry loads for maize, and number of tubers in hundreds for cocoyam.

Total value/month = $47,500, while total value of the food items produced on annual basis is $997,500.00.

Table 31 A comparison of different land acquisition systems between 1978 and 1998.

Land acquisition system	1978 (%) respondents	1998 (%) respondents
Government allocation	70.00	35.56
First occupancy	5.56	14.44
Communal allocation	13.33	32.22
Freeholding	3.33	3.33
Inheritance	7.78	3.33
Rentage	–	10.00
Others	–	1.11

Table 32 Monthly quantities and monetary values of four major products that are derived by village communities from the forest.

Nonwood forest product	Quantity (kg)	Monetary value (US$)
Fuelwood	42,351	818
Leaves	2798	218
Bushmeat	53,900	778
Snails	1452	265
Total/month		= US$2079.00
Total/year		= US$24,948.00

Table 33 Farm products and their monthly values.

Crop items	Monetary value (US$)
Yam	21,200
Cassava	11,100
Plantain	7700
Maize	1600
Vegetable	800
Others	2000
(Beans, rice, etc.)	3100

Revenue generated by Edo State Government from timber felling and logging operations, felling permits, and taungya land allocations to farmers resident within the reserve is presented for 1990–94 on Table 34.

The study clearly demonstrated among other facts that land acquisition for farming purposes was a principal objective of settlements of local communities in the forest reserve. This is in response to the high population density in the southwestern part of Nigeria and the consequent demand for farming particularly in the past two decades.

Furthermore, the study showed that nonwood forest products such as bushmeat, fuelwood, and leaves are basic for their livelihood and their economy.

Table 34 Annual revenue (1990–94) generated by Edo State Government from logging operations, felling permits, and land allocations.

Year	Revenue (US$)
1990	8476.00
1991	8891.50
1992	13,597.23
1993	21,857.80
1994	24,738.21

The importance of medicinal plants and the cultural and medicinal values of wildlife by-products are however discussed elsewhere in this chapter.

Plant biodiversity study

Plant biodiversity survey of Okomu Forest Reserve was carried out by Oguntala and Soladoye (1996) of FRIN. Three major land use zones were sampled in the reserve. These were (a) the permanent sample plot (PSP) within the wildlife sanctuary, representing a most undisturbed part of the reserve; (b) fallow land of more than 5 years; and (c) agroforestry plot currently carrying both forestry and agricultural crops. The PSP represented the natural relatively undisturbed forest presenting the ideal status of plant biodiversity of the reserve. This was used as the standard for comparison of plant biodiversity changes along other land-use forms. Both the fallow plots and the agroforestry plots were located in the buffer zones of the Okomu Forest Reserve. In each land use type, plant biodiversity survey was carried out using 0.2 ha plots that have been established at the center of each land use.

Plant diversity

The PSP contains 44 families represented by 127 plant species made of 1740 stems. In the fallow plot, there are 43 families represented by 90 plants species made of 3995 stems. The agroforestry plot has 28 trees.

The number of shrubs found in the PSPs is 34, while 32 were recorded in the fallow plot. The agroforestry plot had 23 shrubs.

The result of plant biodiversity study (Table 35) thus revealed the following unique characteristics:

The PSPs exhibited the normal structure and physiognomy of the tropical rain forest, while the fallow plot showed some features of secondary forest with the dominant species being *Chromolaena odorata*. The agroforestry plot is dominated by herbaceous, light-demanding seedlings with some arable tree crops.

Table 35 Shannon's and Simpson's similarity indexes for the plant families present in the three land use types in Okomu Forest Reserve.

		Similarity indexes	
No.	**Land use type**	**Shannon's values**	**Simpson's values**
1.	PSP and fallow plots	0.50	0.67
2.	PSP and agroforestry	0.41	0.57
3.	Fallow and agroforestry	0.52	0.70

The families with the highest number of stems varied among the three land use types. For example, in the PSP, the dominant families are Rubiaceae, Violaceae, Commelinaceae, Araceae, Olacaceae, Papilionaceae, Ebenaceae, Euphorbiaceae, Apocynaceae, Meliaceae, Annonaceae, Pandaceae, Ulmaceae, Tiliaceae, Melastomaceae, Caesalpiniaceae, Menispermaceae, Lomariopsidae, Connaraceae, and Icacinaceae.

Whereas in the fallow plot, the dominant families were different with few similarities. These are made up of Compositae, Gramineae, Commelinaceae, Verbenaceae, Pandaceae, Papilionaceae, Agavaceae, Violaceae, Moraceae, Mimosaceae, Araceae, Combretaceae, Caesalpiniaceae, Musaceae, and Rubiaceae. The agroforestry plot was dominated by entirely different plant families with one to three similar families.

The results indicate the presence of a collection of many kinds of habitats that are rich in unique species of plants. This is strongly influenced by human activities mainly on the periphery of the forest reserve. With the current high rate of timber exploitation at Okomu Forest Reserve and its conversion to other land use types, only a few of the adult remnants of a once thriving forest will remain. This situation calls for a tougher control on forest exploitation. Increasing forest clearance and conversion to monoculture agroecosystems will not encourage biodiversity development. Therefore, urgent efforts should be made to conserve the original forest ecosystems and the fallow agroecological systems.

Medicinal plants

A study was also carried out by the FRIN on ethnobotany of Okomu Forest Reserve. Emphasis was placed on ethnomedicine as practiced by the communities around the forest reserve. This is because of the growing importance of traditional medicinal plants in the health-care delivery system of the people of Southern Nigeria. The following account is based on the work of Oguntala and Soladoye (1996).

Plants were collected from the PSP of FRIN, fallow farmland, and agroforestry plots, that is, in areas where research and investigations were also carried out on plant biodiversity. Interviews were held in four village communities on the use of local plants for medicinal purpose. Plant collections were often displayed in villages, thus allowing people of all ages to identify various plants for traditional medicine. Information on 90 plant species belonging to 46 families was obtained (Tables 36 and 37).

Table 36 Some plants of ethnobotanical significance at Okomu Forest Reserve.

S/N	Botanical names	Family	Ailment/uses	Part used
1.	*Asystasia gangetica* (Linn.) T. anders	Acanthaceae	Eczema	Leaves
2.	*Brillantaisia lamium* (Nees) Benth	Acanthaceae	Pains under labor	Root
3.	*Spondias mombin* Linn.	Anacardiaceae	(a) Stomach ache (b) Pneumonia	(a) Bark (b) Bark
4.	*Enantia chlorantha Oliv.*	Annonaceae	Fever	Bark
5.	*Xylopia aethiopica* (Dunal)	Annonaceae	Spices	Fruit
6.	*Alstonia boonei* De Wild	Apocynaceae	(a) Fever (b) Boil (c) Deep cut	(a) Bark (b) Bark latex (c) Stem latex (d) Stem
7.	*Funtumia elastic* (Preuss) Stapf.	Apocynaceae	(a) Dysentery (b) Gum (c) Bird's catcher	(a) Young leaves (b) Latex (c) Latex
8.	*Rauvolfia vomitoria* Afzel	Apocynaceae	Infertility	Root bark
9.	*Tabernaemontana pachysiphon* Stapf	Apocynaceae	Hypertension	Latex
10.	*Anchomanes difformis* (BI.) Engl.	Araceae	Eye diseases	Rhizome
11.	*Colocasia esculenta* (Linn.) Schott	Araceae	Labor pains under delivery of new baby	Rhizome
12.	*Culcasia saxatilis* A. Chev.	Araceae	Wound	Plant
13.	*Secamone afzelii* (Schultes) K. Schum.	Asclepiadaceae	Galactagogue	Leaves
14.	*Ageratum conyzoides* Linn.	Asteraceae (Compositae)	Eye diseases	Leaves
15.	*Chromolaena odorata* (Linn.) King & Robinson Syn. *Eupatorium Odorctum* Linn.	Asteraceae (Compositae)	(a) Malaria (b) Hemostatic agent	(a) Branch (b) Leaves

Continued

Table 36 Some plants of ethnobotanical significance at Okomu Forest Reserve.—cont'd

S/N	Botanical names	Family	Ailment/uses	Part used
16.	*Synedrella nodiflora* (Linn.) Gaertn.	Asteraceae (Compositae)	Ulcer wound	Leaves
17.	*Vernonia amygdalina* Del.	Asteraceae (Compositae)	(a) Soup (b) Stomach trouble (c) Concussion	(a) Leaves (b) Leaves (c) Leaves
18.	*Kigelia Africana* (Lam.) Benth	Bignoniaceae	Chicken pox	Root
19.	*Newbouldia laevis* (P. Beauv.) Seemann ex Bureau	Bignoniaceae	(a) Dysentery (b) Heritage program (c) Boundary	(a) Leaves (b) Branch (c) Plant (d) Stem
20.	*Spathodea campanulata* P. Beauv.	Bignoniaceae	High fever	Leaves
21.	*Ceiba Pentandra* (Linn.) Gaertn.	Bombacaceae	Soap	(a) Bark (b) Stem
22.	*Buchholzia coriacea* Engl.	Capparaceae	(a) Edible (b) Appendicitis	(a) Fruit (b) Leaves (c) Stem
23.	*Combretum racemosum* P. Beauv.	Combretaceae	(a) Seasoner (b) Toothache	(a) Leaves (b) Latex
24.	*Terminalia superba* Engl. & Diels	Combretaceae	Malaria	(a) Bark (b) Stem
25.	*Palisota hirsuta* (Thumb) K. Schum	Commelinaceae	Malaria	Plant
26.	*Momordica charantia* (Linn.)	Cucurbitaceae	Eye diseases	(a) Leaves
27.	*Diospyros piscatorial* Gurke	Ebenaceae	(a) Chew stick (b) Carving	(a) Twig (b) Wood

28.	*Alchornea cordifolia* (Schum. & Thonn.) Muell. Arg.	Euphorbiaceae	(a) Dysentery, stomach trouble (b) Athletic foot	(a) Leaves (b) Leaves
29.	*Alchornea laxiflora* (Benth.) Pax & K. Hoffm.	Euphorbiaceae	Hemorrhoids (Pile)	Root bark
30.	*Bridelia ferruginea* Benth.	Euphorbiaceae	Stomach trouble	Bark
31.	*Jatropha curcas* Linn.	Euphorbiaceae	(a) Stops bleeding (b) Toothache	(a) Latex (b) Stick or twig
32.	*Jatropha gossypilolia* Linn.	Euphorbiaceae	Gonorrhea and abortion	Leaves
33.	*Phyllanthus muelerianus* (O. Ktze) Exell	Euphorbiaceae	(a) Stomach trouble (b) Stop stooling	(a) Root (b) Leaves
34.	*Ricinodendron heudelotii* (Baill) Heckel Subsp. Africanum (Muell. Arg.) J. Leonard	Euphorbiaceae	Stops bleeding	Leaves
35.	*Allanblackia floribunda* Oliv.	Guttiferae	Purgative	Fruit
36.	*Garcinia kola* Heckel	Guttiferae	(a) Cough (b) Edible	(a) Seed (b) Seed (c) Stem
37.	*Harungana madagascariensis* Lam. Ex Poir	Hypericaceae	(a) Painting (b) Swollen baby	(a) Leaves (b) Stem (c) Plant
38.	*Icacina trichantha* Oliv.	Icacinaceae	(a) Edible (b) Fattening	(a) Fruit (b) Leaves
39.	*Solenostemon monostachyus* (P. Beauv.) Briq.	Lamiaceae	Worm expellant	Leaves

Continued

Table 36 Some plants of ethnobotanical significance at Okomu Forest Reserve.—cont'd

S/N	Botanical names	Family	Ailment/uses	Part used
40.	*Ocimum gratissimum* Linn.	Lamiaceae (Labiatae)	(a) Cough (b) Spices (c) Stomachache	(a) Leaves (b) Leaves (c) Leaves
41.	*Peterisianthus macrocarpus* (P. Beauv.) Liben Syn. *Combretodendron africanum* (Welw.). Exel		Cough	Bark or leaves
42.	*Anthonotha macrophylla* P. Beauv.	Leguminosae Caesalpinioideae		
43.	*Distemonanthus benthamianus* Baill.	Leguminosae-Caesalpinioideae	Tonic for pregnant woman	Bark
44.	*Mezoneuron benthamianum* Baill.	Leguminosae-Caesalpinioideae	Craw-craw	Flower and not bark
45.	*Albizia zygia* (DC) J.F. Macbr	Leguminosae-Mimosoideae	(a) Cold in children (b) Increase saliva	(a) Leaves (b) Stick/twig
46.	*Cylicodiscus gabunensis* Harm	Leguminosae-Mimosoideae	Evil spirit expellant	Leaves
47.	*Amphimas pterocarpoides* Harms	Leguminosae-Papilionoideae	Leprosy	(a) Bark (b) Stem
48.	*Baphia nitida* lodd.	Leguminosae-Papilionoideae	High blood pressure	Leaves
49.	*Baphia laurifolia* Baill.	Leguminosae-Papilionoideae		
50.	*Millettia aboensis* (Hook. F.)	Leguminosae-Papilionoideae	Ring worm	Leaves
51.	*Hibiscus surattensis* Linn.	Malvaceae	Bite	Leaves

52.	*Sida acuta subsp.* Acuta *Blumea Syn.* S. acuta Burm. F.	Malvaceae	Wound ulcer	Leaves
53.	*Sida garckeana* Polakowsky Syn. *S. corymbosa* R.E. Fries	Malvaceae	(a) Sperm production (b) Yellow fever	(a) Leaves (b) Leaves
54.	*Khaya ivorensis A.* Chev	Meliaceae	Malaria	(a) Bark (b) Bark
55.	*Trichilia monadelpha* (Thonn.) J.J. De Wilde	Meliaceae	(a) Malaria: venereal diseases (b) Poison expellant	(a) Bark (b) Bark
56.	*Trichilia prieureana A.* Juss. Subsp. prieuriana	Meliaceae	(a) Swollen leg	(a) Leaves (b) Stem Root
57.	*Sphenocentrum jollyanum* Pierre	Menispermaceae	Dysentery, sex stimulant	Root
58.	*Synclisia scabrida* Miers	Menispermaceae	Obesity children	Leaves
59.	*Ficus exasperata* Vahl.	Moraceae	(a) Obesity in children (b) Stops bleeding (c) Stomach trouble (d) Sandpaper	(a) Plant (b) Latex (c) Root (d) Leaves
60.	*Milicia excels* (Welw.) C.C. Berg. Syn. *Chlorophora excelsa* (Welw.) Benth	Moraceae	Malaria, stomach pain	(a) Bark (b) Stem
61.	*Musanga cecropioides* R. Br. Ex. Tedlie	Moraceae	Cough	Latex root
62.	*Myrianthus arboreus* P. Beauv.	Moraceae	(a) Vegetable (b) Antisnake	(a) Leaves (b) Root
63.	*Musa paradisiacal* L.	Musaceae	(a) Eye diseases (b) Edible	(a) Pitch (b) Fruit

Continued

Table 36 Some plants of ethnobotanical significance at Okomu Forest Reserve.—cont'd

S/N	Botanical names	Family	Ailment/uses	Part used
64.	*Pycnanthus angolensis* (Welw.) Warb.	Myristicaceae	(a) Sore threat	(a) Latex (b) Stem
65.	*Boerhavia diffusa* Linn.	Nyctaginaceae	Antiabortion	Root
66.	*Ongokea gore* (Hua) Pierre	Olacaceae	(a) Stomach upset	(a) Bark (b) Stem
67.	*Microdesmis puberula*	Pandaceae	Goat husbandry	Leaves
68.	*Barteria fistulosa* Mast.	Passifloraceae	Protection against witches and wizards	Leaves
69.	*Petiveria alliacea* Linn.	Phytolaccaceae	Madness and poison	Leaves
70.	*Piper guineense* Schum	Piperaceae	(a) Newborn baby (b) Additives	(a) Leaves and seeds (b) Seeds
71.	*Piper umbellatum* Linn.	Piperaceae	(a) Edible (b) Rheumatism	(a) Leaves (b) Root
72.	*Maesopsis eminii* Engl.	Rhamnaceae	Gonorrhea	(a) Bark (b) Stem
73.	*Nauclea diderrichii* (De Wild & Th. Dur.) (Merrill)	Rubiaceae	(a) High fever	(a) Bark only (b) Bark and leaves
74.	*Nauclea latifolia*	Rubiaceae	Temperature control in children	Leaves
75.	*Zanthoxylum gilletii* (De Wild.) Waterman Syn. *Fagara macrophylla* Engl.	Rutaceae	(a) Rheumatism (b) Antimiscarriage	(a) Root bark (b) Bark
76.	*Zanthoxylum zanthoxyloides* (Lam.) Zepern. & Timler	Rutaceae	Infertility	Root bark

77.	*Lecaniodiscus cupanioides* Planch. ex Benth	Sapindaceae	Heartburn	Leaves
78.	*Scoparia dulcis* Linn.	Scrophulariaceae	Piles	Plant
79.	*Datura metel* Linn.	Solanaceae	High blood pressure	Plant
80.	*Solanum torvum* Sw.	Solanaceae	Swollen wound	Seed
81.	*Sterculia tragacantha* Lindl.	Sterculiaceae	Soup	Leaves
82.	*Trema orientalis* (Linn.) Blume Syn. *Trema guineensis* (Schum. & Thonn.) Ficalko	Ulmaceae	Cough	Leaves
83.	*Fleurya aestuans* (Linn.) Gaud. Ex. Miq.	Urticaceae	Skin diseases	Plant
84.	*Clerodendrum capitatum* (Wild) Schum. & Thonn.	Verbenaceae	(a) Pile (b) Ornamental	(a) Branch (b) Plant
85.	*Stachytarpheta cayennensis* (L.C. Rich.) Schau	Verbenaceae	Malaria	Plant
86.	*Aframomum melegueta* Schum.	Zingiberaceae	(a) Pepper soup	(a) Seeds
87.	*Costus afer* Ker-Gawl	Zingiberaceae	(a) Sugarcane (b) Washing	(a) Stem (b) Leaves
88.	*Zingiber officinale* Rosc.	Zingiberaceae	Malaria	Leaves

B, benin; *A*, agroforestry plot; *S/T*, small tree; *Es*, Esan; *F*, fallow land plot; *M/T*, medium tree; *H*, herb; *P*, permanent sample plot (PSP); *L*, liana; *S*, shrub; *Ho*, homestead; *C*, climber; *T*, tree.

Table 37 Plant parts used in ethnobotany.

Part used	Number of cases recorded
Latex	10
Leaves	53
Root	13
Fruit/seed	12
Bark	20
Rhizome	2
Whole plant	11
Twig/branch	6
Wood	16
Flower	1
Pitch	1

Table 38 Occurrences of plants in land use forms of Okomu.

Habit plot	Permanent sample plot	Agroforestry	Fallow homestead plot	
Tree	22	20	19	13
Shrub	1	7	5	6
Herb	4	10	10	9
Climber	2	1	0	0
Liane	0	1	1	0

The survey further showed that communities have a wide knowledge of medicinal plants for malaria fever, followed by children and pregnant women's ailments. These herbal preparations were readily available in their first aid shelves. The malaria medication was often prepared with local gin and taken frequently than any other medicine. The high rate of malaria sickness was probably due to their nearness to the forest that harbors a lot of mosquitoes.

This study has revealed the richness of Okomu Forest Reserve trado-medicine flora and further demonstrates the fire need to conserve the tropical high forest ecosystems for human welfare Tables 36 and 38.

Wildlife conservation and production of bushmeat

Officials of the NCF, an NGO, and some international staff with those of Edo State Government Forestry Department carried out the inventory and monitoring activities of the wildlife resources of the reserve in 1987.

Because of the very dense tropical rain forest vegetation, direct total counts by aerial and ground census was not feasible. Therefore all population counts were done by indirect assessments using the standard King's Strip Census (KSC) and Timed Species Counts (TSC) methods. One-kilometer transects were taken at random locations at right angles to the major roads, and observations were made on animals and their activities, including droppings and defecations, calls, and animal tracts.

Population inventory was at a compartment considered to be an average between high population and low population compartments. The results were then extrapolated for other compartments and presented on Table 39. Current monetary values were obtained for each species by pricing live animals and their carcasses in villages and towns around the forest reserve. These prices were then multiplied by total population estimate for each species to obtain estimated values for each species and for most of the wildlife species

Table 39 Total wildlife population estimates and their money values at Okomu Forest Reserve.

Species	Average price/ individual animal (US$)	Quantities (for 322 compartments)
Duiker—two species, Yellow-backed (*Cephalophus silvicultor*), and black duiker (*Cephalophus niger*)	3.1	57,960
Gray duiker (*Sylvicapra grimmia*)	57.5	6440
Forest buffalo (*Syncerops caffer*)	1.9	167,440
Civet cats (*Viverra civetta*)	1.1	57,960
Genet (*Genetta pardina*)	18.8	1610
Leopard (*Panthera pardus*)	2.5	128,800
White-throated monkey (*Cerocopithecus erythrogaster*)	2.5	193,200
Mona monkey (*Cercopithecus mona*)	2.5	2,382,800
Putty-nosed guenon (*Cepcopithecus nictitans*)	62.5	–
Elephant (*Loxodouta africana cyclotis*)	3.1.	80,500
Bushbuck (*Traelaphus scriptus*)	1.5	644,000
Cane rat (*Thryonomys swinderianus*)	1.5.	193,200
Porcupine (*Atherurus* sp.)	0.6	225,400
Giant rat (*Cricetomys emini*)	0.6	96,600
Pangolin (*Manis triscuspis*)	0.3	64,400
Ground squirrel (*Xerus* sp.)	0.4	38,640
Flying squirrel (*Anomalurus* sp.)	0.3	64,400

Continued

Table 39 Total wildlife population estimates and their money values at Okomu Forest Reserve.—cont'd

Species	Average price/ individual animal (US$)	Quantities (for 322 compartments)
Potto (*Perodctritus potto*)	1.9	80,500
Redcap (*Cerocebus galeritus*)	0.5	579,600
Mongoose (*Herpestes naso*)	2.8	8050
Chimpanzee (*Pan troglodytes*)	3.3	
Crocodile: Long-nosed (*Crocodylus niloticus*), short-nosed (*Osteolaemus*)		48,300
Galago (dwarf), (*Galagoides demidovii*)	0.3	
Red river hog (*Potarmochoerus porcus*)	5.0	64,400
Monitor lizard (*Venanus niloticus*)	1.0	161,000
Tree hyrax (*Dendrohyrax* sp.)	1.0	225,400
Climbing squirrel (*Zerus* sp.)	0.1	64,400
Tortoise	1.3	96,600
Chameleon	2.0	38,640
Royal python (*Python requinus*)	1.0	32,200
Cobra	0.6	25,760
Green mamba	0.4	64,400
Snail (*Archachatina marginata*)	0.1	129,900
Guinea fowl—two species (*Guttera edourdi* and *Numida meleagris*)		96,600
Birds		96,600
Gray plantain eater (*Crinifer piscator*)		96,600
Hornbill (eight species)—(*Lophocros nasutus*)		96,600
African gray parrot (*Psittacus erithacus*)		
Red-billed wood dove (*Turtur after*)		
Woodpecker (*Mesopicus Geortaue*)		
Eagle (*Falconidae*)		
Owl (*Ptilopsis* sp.)		
Senegal coucal (*Centropus senegalensis*)		
Yellow wagtail (*Budytes flavus*)		
White-throated bee-eater (*Merops albicolis*)		
Francolin (*Francolinus bicalcaratus*)		
Total wildlife surveyed		8,754,700
*Less endangered species		
Total estimated wildlife value for the entire forest reserve		

in the forest reserve. The total values for all species in the forest reserve are presented on Table 39. It shows that in 1987 the forest reserve contains an estimated 8,754,700 animals, valued at about US$7,384,931.3.

Sustainable offtake for bushmeat production

The project management, after the inventory survey, can therefore proceed to make calculations for sustainable offtake of wildlife for bushmeat production, based on the following assumptions:

a. The intrinsic rate of increase of each species of wildlife, which depends on its total population in a particular area, sex ratio, rate of recruitment, and natural mortality pattern.

b. The existence of the sanctuary where all human activities including hunting and logging are prohibited. The sanctuary thus provides a core breeding area for wildlife. Forest compartments contiguous to the sanctuary and the buffer zone are therefore expected to be relatively rich in wildlife.

c. The effectiveness of the management (game protection and monitoring and regulation of logging activities and sustenance of forest integrity) by officers of the NCF and officers of Edo State Forestry Department.

d. Strict adherence to hunting regulations and levels of exploitation of wildlife by licensed hunters under the management of Edo State Forestry Department.

e. Considering the factors (a–d) above, wildlife populations were expected to increase considerably with a high turnover in every 5 years.

Hunting practices for bushmeat production

Hunting permits and regulations

The prospective hunter in the forest reserve who wishes to hunt for bushmeat has to apply to the police for permission to possess firearms and to purchase ammunition. Hunting permits cost about $6.0 per year in 1993 during the last survey, and the application has to undergo some physical and general tests: he must pass an eye test to ensure that the hunter can see sufficiently well so that he does not constitute a risk to others. There will also be an inquiry to see if the applicant has any criminal record. This is to avoid giving a license to armed robbers and criminals. The physical fitness of the prospective hunter and his courage to face wild animals in the bush, his tolerance in dealing with other hunters, and his control under provocation are all aspects of the applicant's behavior and conduct with which the authorities must be satisfied before a permit is issued.

In addition to the rigorous examinations that the prospective hunting has to pass before he is issued a license, he must also observe the following rules: guns must not be loaded at home or anywhere other than at the hunting ground. No hunter may lend his gun to anyone else. A hunter must not fire at an unidentified object or animal. In daytime, fire arm must be angled up or down, not straight ahead (at the height of other humans in the area).

Bush meat production and promotion of economic returns

Hunting tools

Forest mammals are still hunted in the tropical rain forest in Nigeria by simple tools such as snares traps, spears, bows and arrows, poisoned baits, and nets. Dogs are also used in community traditional group hunting. The use of dane guns, shotguns, and rifles came as a result of man's advancement and urbanization and the corresponding higher demand of wild animal protein. However, primitive and modern hunting tools are still employed in Okomu Forest Reserve. Details of hunting techniques are similar to those described for West Africa earlier in this book.

The following are popular taboos and local rules connected with hunting tools among the hunters in Okomu Forest Reserve:

> *If a hunter uses a charm in killing an animal, he will not use it to feed his family [directly]; otherwise he believes that his children will start to die one after the other: instead, he sells the meat to other people.*

> *If an animal is killed by the use of poisoned trap or arrows, all the intestines must be removed before the carcass is used for food.*

According to Darling (1960), hunters in the forest reserve encountered several problems, and the following is a hunter's account of his predicaments: "mosquitoes, hunger, and thirst are all parts of the hunter's lot. Snakes, "leopards," and other wild animals are additional hazards, but accidental shooting by other hunters is possibly the greatest problem. Cases of missing hunters, who are never seen again, are usually announced over the radio." The dense tropical rain forest with its complex ecosystems and a large variety of animals also constitute a lot of hazards to hunters. The following tale was also recorded from a bushmeat hunter by Darling (1977):

> *It was rainy, cold and chilly. I was hunting deep inside the forest when the rain extinguished my headlight. The matches in my pocket were wet and soggy. It was very dark and I could not see my way: I knew I was in for a bad night. I climbed up a tree to avoid dangerous wild animals. Just as I was breathing a sign of relief, a python*

coiled round my neck. I held my breath, but it was not good enough. I jumped, fell down from the tree, badly hurting my legs in the process. I crawled into a corner of the forest and crawled up again, shivering. As I prayed for dawn to come, my exposed skin was badly bitten by mosquitoes.

These are some of the tales of local hunters in the tropical rain forest of West Africa.

Economic and cultural significance of forest wildlife

As part of the survey of economic importance of wildlife in Okomu Forest Reserve, their cultural significance among the local communities was also studied. Apart from the well-documented roles of wildlife in local and international tourism and as a source of bushmeat, local communities in Nigeria and indeed along the West Coast of Africa depend heavily on wildlife by-products for ornamentals and traditional medicines. For example, by-products such as tusks, hooves, horns, beads, feathers, skins, and teeth of wild animals are used in producing highly priced ornamentals. Traditional chiefs and warriors wear leopard skins in ceremonies as a symbol of prowess and supremacy. Python skins are very important dressing components for masquerades in their annual celebrations, while elephant tusks are important to the installation of traditional rulers. Wildlife by-products in traditional medicine is however one of the most significant uses of wildlife among the high forest local communities.

Table 40 shows the results of a recent survey of various tropical high forest wildlife by-products that are used in traditional medicine. These uses range from direct application of by-products for curing certain ailments to spiritual uses for invoking spirits to perform certain functions. Furthermore, recent surveys showed that both medicinal and spiritual uses of wildlife by-products are very popular among local communities. Therefore, it is not a question of whether one believes in these rather obscure uses to which wildlife by-products are put, but it demonstrates clearly how the culture of local communities in tropical high forest is interwoven with their environment and the nonwood forest products.

Management and research priorities

Human activities are principal factors that generally lead to the loss of biodiversity such as deforestation, habitat loss, wildlife scarcity, and erosion. These problems still occur at Okomu Forest Reserve due to heavy human

Table 40 Medical uses of wildlife in some villages around Okomu Forest Reserve.

Animal species	Part of body used	Medical uses
Elephant	Head	It is used to prepare medicine and applied on women who flirt about
		It is also used for a myriad of witchcraft by elders
	Feces	It features in various important traditional medicines
	Bone	Used as "Ajako" and for traditional medicine known as "egbe." This is used in time of danger for easy disappearance
		In preparing traditional medicine and witchcraft, a piece of meat is attached to the bone and mixed with traditional black soap that is used for bathing. This is believed to bring wealth. It is commonly used by businessmen and women
	Skin	It is used for protection and is prepared in the form of talisman that is thought to bring good luck. It is used by hunters for protection against dangerous animals like lions and leopards
		It is also used in witchcraft in the form of padlock. The user enchants on it and lucks it up against his enemy. The curse becomes severe on the enemy, if the padlock is thrown into a river or a pit lavatory
	Footprints	Soils defined by elephant footprints are used medicinally to secure justices in the court of law
Leopard	Skin	It is prepared medicinally with black soap and other ingredients. This may then be used by local boxers and wrestlers that provide them with extra strength to dish out powerful blows that make opponents surrender
		The latter is also used by businessmen and women in bathing, and it is believed that it provides wealth and riches
		It can also be used in padlocking enemies to cause stagnation and disasters in their business and relationships
		It is used to prepare talisman for protection against witchcraft
		It is also used for witchcraft by politicians, civil servants, and chiefs to command respect and to win popularity among working colleagues and voters
	Urine	The urine is used to prepare traditional medicines that are used by night guards. This is used to tie down thieves and make them unconscious
Python	Teeth	Ashes of burnt teeth are mixed with orange lemon, coconut oil that is used together to prepare a concoction for children to cure stomach troubles. It is also used as antisnake bites
	Liver	Used to prepare medicines to cure convulsion in children

Buffalo	Bone	The bone is burnt and mixed with other ingredients to prepare concoctions for prevention against witchcraft
Guinea fowl	Liver + intestines + feathers and skull	The mixture of these with alligator pepper is mused traditionally to cure dizziness
Bushbuck	Bone	Used to prepare concoction for children between 3 months and 3 years of age for the development of strong bones and rigid body. The bones are burnt and mixed with other ingredients that are applied to the legs to make the bones strong: it is also used by pregnant women who find it difficult to walk
Gorilla	Skin	Used to prepare witchcrafts for initiating madness against enemies
Cane rat	Skin	A mixture of the skin, intestine, and black soap are prepared and used to bathe children for general immunity against all disease
Aardvark	Paws	This is used to prepare concoctions to cure infection in finger nails
	Tail	Used by elders to command respect among their subordinates
Green monkey	Skull	Used to prepare concoctions for children between 6 months and 1 year old to prevent and cure stomach ailments
Red river hog	Bone	Used to prepare concoctions to cure teething problems in young children
Chimpanzee	Skin	This is used in traditional medicine by pregnant women to prevent miscarriage
Red-flanked duiker	Skin	Used by blacksmiths and goldsmiths for attracting customers
		The skin is used to cover knife cases, purses, and rings by hunters to protect them from dangerous wild animals
Flying squirrels	Tail	Wrestlers use this against opponents to find it impossible to floor them or put them to submission during a wrestling match
		It is used to prepare medicine with other ingredients to avert miscarriage in pregnant women
Maxwell duiker	Skin	The skin is used by blacksmiths and goldsmiths, and it is believed to bring good luck by attracting customers
	Horn	Special medicines prepared by hunters are believed to be more potent when kept in the horns of the duiker
Red river hog	Bone	The bone is burnt and mixed with shea-butter oil that is applied to the body to prevent boils and other skin diseases
Lion	Heart	Used by warriors for poisoning spears and arrows during war or for killing humans and for hunting purposes
		Used to prepare concoction for children to drink for immunity against diseases

population that is characteristic of southern Nigeria, most of whom depend on rural agrarian economy, which is based on forest products. There is thus an urgent need to evolve a management and legal strategy that will limit human settlements in and around the forest reserve.

Furthermore, the involvement of local communities in this tripartite management system of Okomu Forest Reserve has to be further developed. The management must evolve an institutional and legal framework for community participation in decision-making process of management principles and practice that will enhance forest conservation and resource utilization. Their actual involvement in the day-to-day running of the forest reserve and in sharing the benefits of forest conservation must be clearly defined. Community participation must be developed sustainably along the guiding principles of Communal Area Management Programme for Indigenous Resources (CAMPFIRE) in Zimbabwe, the Administrative Management by Design (ADMADE) in Zambia, and also the Luangwa Integrated Resource Development Programme (LIRDP) in Zambia. What these programs have in common is a culturally viable, realistic administrative design for the involvement of local communities in integrated utilization of their natural resources.

It is also important to discourage any pressure from the government to dereserve more areas of the forest reserve for monoculture of oil palm and rubber tree plantations. Furthermore, there is need to develop sustainable silvicultural program to restock and replenish the trees of the forest reserve, which has been depleted of some of its rare woody plants through illegal and indiscriminate logging. Agroforestry farming system should also be encouraged among the peripheral communities, as this system will enhance soil fertility and maintains tropical rain forest mosaic.

A thorough inventory of the wildlife of the Forest Reserve must be carried out. Wildlife management must be based on detailed knowledge of the varieties of species, their total populations, sex ratio, and other population characteristics upon which viable and sustainable game cropping depends. It is important also that game cropping be systematic and organized for optimum sustainable yield of bushmeat. There must be strict hunting laws and a massive campaign in the environs of the forest reserve to increase public awareness of the importance of wildlife and forest conservation. There must also be a program of research into management practices for sustainable economic benefits to the state government and the rural communities.

CHAPTER 16

Ex situ conservation of wildlife: Domestication of the African giant rat (*Cricetomys gambianus*)

Ex situ literally means "off-site." It is the process of protecting an endangered species, variety, or breed of plant or animal outside of its natural habitat, for example, by removing part of the population from a threatened habitat and placing it in a new location, which may be a wild area or within the care of humans.

The reasons that led to the development of the study reported here are both practical and scientific. The African giant rat (*Cricetomys gambianus*) and the grass cutter (*Thryonomys swinderianus*) have been sources of protein in West Africa from time immemorial. With increasing food shortage, these sources of protein continue to be of interest. The scientific reasons relate to the continued debate on the feasibility of domesticating wildlife, both large and small mammals, for the use of man. Despite this debate, the numbers of instances of serious attempts in recent years to breed wild animals and domesticate them are very small. Thus, in many respects, the study reported here was a pioneering effort by the author to assess the feasibility of domesticating animals that are currently sources of protein in parts of West Africa and in certain areas near forest reserves as major sources of protein.

The underlying philosophy motivating this and similar studies was that it is better to assess all indigenous sources of protein with the introduction and selection of exotic forms. Also, it must be stressed that the problem is of some urgency because of the rapid disintegration of natural ecosystems. Furthermore, Africa is under the pressures of deforestation, farming, and pastoralism. It is hoped that the information provided in this chapter will stimulate other workers to look at other indigenous forms of protein that may be domesticated for the benefit of the local people and the long-term stability of natural ecosystems of the African continent.

Wildlife Conservation in Africa
https://doi.org/10.1016/B978-0-12-816962-9.00016-8

The following species that appeared to be relatively prolific and widely eaten by West Africans were selected for domestication:

The African giant rat (*Cricetomys gambianus*)

The grass cutter (cane rat) (*Thryonomys swinderianus*)

The guinea fowl (*Numida meleagris galeata*)

Of all these species, the African giant rat and the grass cutter have received the greatest attention. As a prerequisite to their domestication, their social acceptance in Southern Nigeria was studied by Ajayi (1975). Investigations showed that 71.4% of the sample population accepted the two species as a source of meat. For those who did not eat giant rat, reasons adduced for its nonacceptance were age-long taboos in different tribes. For examples, in Oyo State of Southwestern Nigeria, it is believed that the giant rat is used for making witchcraft to torment people or to utter an evil wish against an enemy and therefore should not be eaten. In some parts of Ondo State, it is believed that the giant rat was an ally who helped them out of many intertribal wars and therefore should be respected by not being eaten. Furthermore, it is believed that a pregnant woman who eats giant rat would give birth to a rogue and an outcast.

The foundation of wildlife domestication team

The domestication process started in 1972 and continued to date. The author and the following colleagues were responsible for the initial successes recorded in the domestication efforts:

Professor S.S. Ajayi
Professor Ebenezer Agbelusi
Mr. Sola Ayandare
Alhaji Lateef Suberu
Mr. Femi Owolabi
Mr. Maxwell Ibeabuchi (late)
Professor O.O. Tewe
Professor E.O. Faturoti
Dr. J.S.O. Ayeni
Professor G.A. Lameed
Professor O.O. Dipeolu (late)
Professor B.O. Ikede
Professor O. Olowookorun

Rodents as a source of protein

Rodents form a significant portion of the total amount of bushmeat consumed in West Africa today. Asibey (1970) found in one market in Accra during 1 year 73 tons of grass cutters *Thryonomys swinderianus* made of 15,564 animals, which were sold at an equivalent of US$73,000. The annual bushmeat consumption in Ghana was 8486 tons valued at US$7,358,172. There were no data for the giant rat consumption in Ghana, but it is thought that they are as important as grass cutter. In Southern Nigeria, Ashford (1970a,b) reported that the African giant rat is a highly favored rodent. While, Bates (1905) stated that hunters in Southern Cameroon went on distant camping trips for the expressed purpose of seeking and smoking giant rats for meat.

Rosevear (1969) and Booth (1959) stated that giant rats and grass cutters have a wide distribution, occurring everywhere in Africa South of the Sahara. Thus the species is adapted to a wide variation in climate and ecological zones. This kind of adaptation to various environmental conditions is an attribute that Bigalke (1964) recommended as an important criterion in selecting animals for domestication.

Rodents are small mammals that have a higher turnover rate than larger herbivores when the criteria for productivity are considered, such as age at first breeding, gestation period, litter size, length of time between successive births, and growth rates (Talbot et al., 1965). This further suggests that rodents can be managed for short-term production of meat. Rodents are predominantly vegetarian (Rosevear, 1969) and might not necessarily compete with man for food if domesticated.

There is also an evidence that the African giant rat has been kept successfully in captivity as pets (Ashford, 1970a,b).

Scope of the study

The project was aimed at determining the possibility of domesticating the African giant rat and the grass cutter for short-term production of meat and also for increasing man's variety of meat from domestic sources (Ajayi, 1974). However, the two species will be treated separately in two different chapters as examples of ex situ conservation of African wildlife. The study examined the characteristics of its habitats in order to understand its requirements in domestication. Other items of investigation included:

a. A survey of social acceptance of giant rat for consumption

b. Some aspects of the ecology and burrow patterns as prerequisites for the domestication of the giant rat

c. Disease surveillance of the wild rats to find out the veterinary implications of domesticating this rodent from the point of view of public health

d. Carcass analysis in order to know its meat production potential
e. Cage designs in order to house rats successfully in captivity
f. Reproductive performance
g. Growth rates

However, it is only the aspects of their biology that were essential to the domestication process that is reported here.

Domestication process

Published information on wild animal husbandry has shown that the management of small animals in captivity, particularly rodents, in zoos and in laboratories, was usually based on caging. Ajayi (1974) gave an account of eight species of rodents of the suborder *Hystricomorpha*, which thrived well in cages. The tuco-tuco *Ctenomys talarum* of South American origin is strictly subterranean in its native habitat but soon became adapted to life in wooden cages. The plains viscacha *Lagostomus maximus*, also a South American rodent that weighs 3–8 kg and that digs extensive burrow systems had bred successfully when placed in pairs in wooden cages. The acouchi *Myoprocta pratti* lives a solitary life in its natural habitat in the undergrowth of forest regions of the West Indies and northern South America but has bred successfully in wooden cages. Wild guinea pigs *Cavia aperea*, the cuis *Galea musteloides*, and the degu *Octodon degu* all natives of South America have been reared successfully in cages. Pogosianz (1972) also reported taming and breeding the steppe lemming *Lagurus* spp. in cages. Gambino et al. (1965) reported the breeding success of the little pocket mouse *Perognathus longimembris* even though it has a solitary habit and is aggressive. Egoscue (1972) reared the Polynesian rat *Rattus exulans* in captivity although it is strictly monogamous in nature. Egoscue (1972) also bred the East-Central African long-tailed small-pouched rat *Beamys hindei* in metal and wooden cages, even though in nature, it is nocturnal and burrows extensively.

The husbandry of other small laboratory mammals such as the rabbit *Oryctolagus cuniculus*, the laboratory rat that is a descendant of the brown rat *Rattus norvegicus*, and the guinea pig *Cavia porcellus* also involved the use of cages.

Other factors considered in caging small mammals

The foregoing accounts of the past attempts to manage wild rodents under laboratory conditions showed that these animals can adjust to the abnormal environment of captivity. It also indicates the success in rearing various species of rodents in cages. However, UFAW (1971) recommended that in order to design suitable cages for rearing wild small mammals in the laboratory, the behavior pattern and habitat requirements of the species concerned must be carefully considered. For example, wild rats are nocturnal in nature, and their laboratory descendants have accordingly shown preference for solid-walled cages. Pogosianz (1972) found that steppe lemmings have a high metabolic rate and suggested that when placed in captivity, care should be taken to ensure that they have adequate ventilation. Lane-Petter (1972) noted that rats are highly susceptible to respiratory diseases and suggested that ventilation may be a very important factor in their environment. Redfern and Rowe et al. (1964) and Lane-Petter (1972) found relative humidity to be important to rats. If the relative humidity falls below 45% and the temperature is low, outbreak of ring tail disease usually occurred. A relative humidity of about 50%–60% was suggested for the maintenance of laboratory rats.

UFAW (1971) and Egoscue (1972) also suggested an ambient temperature of 20–26°C for rodents in captivity. They found a temperature of about 36–38°C to be fatal to adult steppe lemming and laboratory rats although the optimal temperature for newborn animals was 35–38°C. Space requirement could also be important in rearing wild animals in captivity. UFAW (1971) have found that disease outbreaks among rabbits are more common when they are crowded. The recommended floor space for caging a pair of small breeds was 61 × 76 cm, medium breeds 61 × 91 cm, and 61 × 122 cm for larger breeds.

Some aspects of the ecology of the African giant rat that are considered relevant to its domestication

The African giant rat is found in a wide variety of habitats in Ibadan district of Southern Nigeria. These include an abandoned farm in a degraded deciduous forest; a zoological garden complex in Agodi; and a forest plantation of teak (*Tectona grandis* L.F.), human habitation, and derived Guinea savanna. They located their burrows in the most heavily shaded parts of the habitants. Burrows were found mostly inside deserted termite mounds and among roots of trees probably to provide insulation from heat. They are nocturnal, omnivorous rodents, feeding on a wide variety of root crops, fruits, mollusks, and insects. Fruits of the oil palm (*Elaeis guineensis* Jacq.) or their remains were found in all burrows. Information on habitat preferences, food habits, and that obtained from the dimensions and design of burrows that are reported in this chapter later provided a basis for the design of cages and husbandry techniques for a program of domestication of this rodent, which was carried out at University of Ibadan, Nigeria, by the author and his wildlife domestication team.

The previous observations available on its habitats are conflicting. Bates (1905) reported that these rats are found in high forest rather than in urban areas, and Sanderson (1940) has found that within the forests, giant rats take up their abode among the roots of large trees and in holes in large rotten and fallen tree trunks. On the other hand, Rosevear (1969) believed that giant rats are semidomestic, as they are found near and occasionally inside houses. They have also been found to inhabit termite mounds, along river banks in grassland or among rocks.

Burrow patterns

Giant rats live in borrows that are provided with exit holes (Booth, 1959). Rosevear (1969) described these burrows as not only simple holes but also usually an entrance leading to a system of galleries with side branches and separate chambers for food storage, droppings, and sleeping. According to Rosevear (1969), giant rats live a solitary life, each occupying a burrow except at breeding or nursing time. The entrance shaft of the burrow is usually sealed with chewed earth and sometimes reinforced by incorporation of sticks or even metals where burrows are located close to human habitation. It has also been reported that giant rats are omnivorous, eating everything from termites and snails to palm nuts (Hatt and Reiss, 1940; Rosevear, 1969).

The study areas

An abandoned farm located in a deciduous forest at the International Institute of Tropical Agriculture (IITA), Ibadan, Nigeria: This study area was about 8 km along the Ibadan-Oyo road. The abandoned farm was about 5 ha lying northwest of a reservoir and was bisected by a stream whose swampy sides covered more than two-thirds of the area. The study area was bounded on the south by a cocoa *Theobroma cacao* plantation. The natural vegetation included dominant trees such as *Spathodea campanulata*, *Ricinodendron heudelotii*, *Trema guineensis*, *Ceiba pentandra*, *Alstonia boonei*, and *Eleaeis guineensis*. There was no hunting in this area since 1967 when the farm was abandoned.

A zoological garden complex, Ibadan, Nigeria: The area was made up of a zoological garden, a teak (*Tectona grandis*) plantation and an abandoned farm, and situated northwest of Ibadan city was prohibited in the area.

The abandoned farm had abundant cassava (*Manihot esculenta*), maize (*Zea mays*), and oil palm trees (*Elaeis guineensis*), when the study was conducted.

A derived Guinea savanna near Ibadan, Nigeria: The area was about 5 ha southwest of Ibadan and dominated by such tree species as *Uapaca togoensis*, *Ceiba pentandra*, *Anonidium mannii*, *Chrysophyllum albidum*, *Bosques angolensis*, and *Eleaies guineensis*.

Studies on burrow patterns of giant rats at University of Ibadan Halls of Residence, Nigeria

This represented a human dwelling habitat within a deciduous rain forest area of southwestern Nigeria. The following observations were made on the burrow patterns of giant rats in the study areas:

a. Location of burrows in relation to trees, topography, and termite mounds
b. Dimensions of burrows; distance between entrance and nesting chambers, entrance and exit holes, size nesting chambers, diameter of entrance, depth of burrows, and also the total number of exit holes per burrow were all studied at the abandoned farm site
c. Design of burrows

Food habits

The food habits of giant rats were studied by stomach content analysis of ten rats killed at the zoological garden complex study area. This method has been used by Day and Linn (1972) for feral mink in England. Elley (1970) for Cape hare in Kenya and Field (1972) for wild ungulates in Uganda. In addition, food remains were examined in the nesting chambers of the excavated burrows at the zoological garden complex in Ibadan. Quantitative presentation of results was by weight of each food item in the burrows expressed as a percentage of total weight of all food item found.

Results

Location of burrows: The distribution of burrows on the abandoned farm is shown in Tables 41 and 42. About one-third of the burrows were found inside deserted termite mounds and another one-third underneath roots of trees. Highest concentration of burrows coincided with dense tree and ground covers (Table 43 and Figs. 20 and 21).

Most burrow systems showed a basic pattern of a nesting chamber, one or more exit holes, and a system of foraging burrows. However, the extent

Table 41 Site distribution of burrows of giant rats at the International Institute of Tropical Agriculture (IITA), Ibadan.

Site of burrows	Termite mounds	Other mounds	Foot of palm trees	Foot of other live trees	Foot of dead trees	On level ground	Other areas
No. of burrows	15	6	2	6	2	6	8
% of total number of burrows $n=45$	33.3	13.3	4.4	13.3	4.4	13.6	17.7

Table 42 Dimensions of giant rate borrows at the International Institute of Tropical Agriculture (IITA), Ibadan.

Range of depth of burrows (m)	0.3–0.1
Range of length of burrows (i.e., between the entrance and nearest exist (m)	0.9–2.9
Burrows less than 1–8 m long (%)	50
Burrows of diameter of entrance (cm)	0.9–20.3
Burrows with diameter of entrance 15 cm or less (%)	50

Table 43 The distribution of nesting chamber and exit holes in giant rat burrows at the International Institute of Tropical Agriculture (IITA), Ibadan.

No. of exists per burrow	% burrows (total no. = 45)
0	4.4
1	48.9
2	28.9
3	15.6
4	2.2
No. of nesting chambers per burrow	% burrows (total no. = 45)
1	85.2
2	14.8

and shape of these burrows varied considerably. The different designs of burrows observed at the zoological garden complex, Agodi, Ibadan (Fig. 22), may be grouped within six basic designs:

a. Those burrows located inside deserted termite mounds. Rats simply smoothed the walls of the original cavity and utilized termite tunnels for their burrows. Burrows did not usually extend beyond the termite mound. Exit holes were blocked with chewed earth.

Fig. 20 Burrow of the African giant rat through a concrete foundation at a students' hall of residence, University of Ibadan.

Fig. 21 Burrow of the African giant rat in a deserted termite mound.

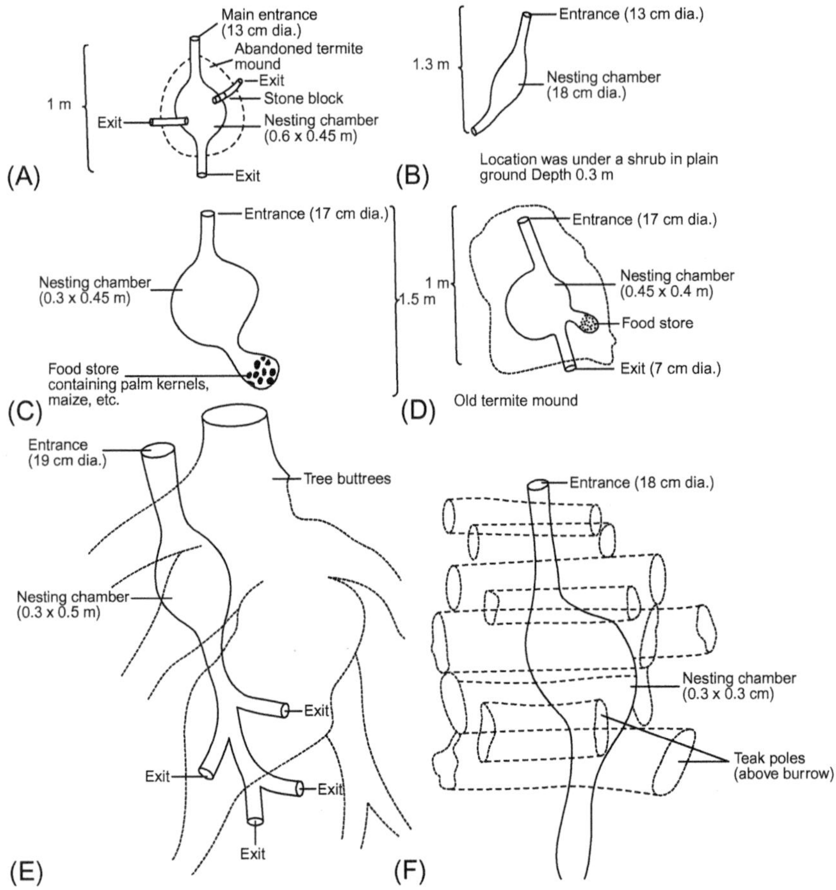

Fig. 22 Designs of burrows of the African giant rats observed at the Agodi Garden Ibadan, Nigeria.

b. The burrows that were located on plain ground in shrub land were simple, almost assuming a straight-line design. Such burrows usually had a single exit hole.

c. Burrows found among roots of teak characteristically had a large nesting chamber and no exit holes. The roots provided shelter, insulation, and firm noncollapsible walls for the burrows.

d. One burrow consisted of a large nesting chamber and a separate food store. The burrow was also located on a termite mound, but the nesting chamber was much cleaner than in any other burrows observed.

e. Some burrows were found among the roots of large dead trees. The entrances were large and clean, leading to one or more nesting chambers that again led into a gallery of holes that terminated in several exit holes. Exits were partially sealed with earth and dried leaves, when rats were in residence.

f. One burrow was located underneath a pile of felled thick poles. The design was simple and consisted of an entrance that led to a large nesting chamber but no exit hole. This was an unoccupied burrow. Its simplicity, with little sign of activity by the giant rat, showed that it might have served a temporary purpose.

The nesting chamber of the giant rats was more than a simple sleeping area and contained in addition to the nest proper, a food store, and a sanitary area. An age gradient of the contents of the chamber was easily apparent. Close to the entrance of the chamber were nesting materials that consisted mainly of dry grass and leaves; next was an area packed with progressively older mixtures of food materials, sticks, and stones; and further back was a mixture of decomposing food materials and feces. The chamber of an occupied burrow was often warm and invariably contained *Hermimerus talpoides* (an insect ectoparasite of the rats) at different stage of their development.

While excavating the giant rat burrows, usually starting from the entrance, rats characteristically blocked the burrows halfway between the entrance and the nesting chambers. The blockade was made of earth, stones, leaves, and sticks. It was so firm that an inexperienced hunter of this rodent would abandon the digging with a false assumption that the burrow had terminated. It was common to encounter two such blockades while excavating an occupied burrow. On certain occasions when all the exit holes had not been identified and blocked before digging, rats used the exit holes to escape.

Food habits

The following were important food items found in the stomachs and nesting chambers of giant rat burrows: cassava, *Manihot utilissima* Pohl; Chinese yam *Dioscorea rotundata*, Poir; Chinese yam, *Discorea esculenta*, Lour; sweet potatoes, *Ipomoea batatas*, (Linn) Lam; maize *Zea mays* Linn; pawpaw, *Carica papaya*, Linn; banana, *Musa* spp; and *cowpea*, *Vigna unguiculata*, (Linn.) walp. Fruits of the oil palm *Elaeis guineensis* Jacq. and their remains were found in all burrows. Tables 44 and 45 give the result of the stomach content analysis and components of food stored in the nesting chambers of burrows.

Location of burrow was inside a vacated termite mound. Maximum depth about 1 m.

Location was under a shrub on plain ground. Depth 0.3 m.

Table 44 Mean stomach contents of giant rats trapped at the zoological garden, Agodi, Ibadan.

Food items	% stomach content	
	Males (n = 5)	Females (n = 5)
Palm fruits	72.0	16.5
Fruits other than palm fruits	1.75	22.8
Vegetables	1.65	33.0
Tubers (yams)	–	15.5
Insects	7.0	6.0
Seeds	12.2	5.0
Miscellaneous	4.4	2.0
Total %	100.0	100.0

Table 45 Analysis of food stores in occupied giant rat burrows at the zoological garden, Agodi, Ibadan.

Food items	% contents of burrows	
	Males (n = 5)	Females (n = 7)
Palm fruits	48	52.5
Tubers (yams)	3	17.5
Seeds	43	20.0
Bones	–	5.8
Miscellaneous (unidentifiable items)	6	4.2
Total %	100.0	100.0

The nesting chamber contained leaves, loose sand, dry grass, and barks of trees that are used as a bed. Burrows of this kind were found underneath roots of teak. N.B. There is no exit to the burrow.

Location inside a vacated termite mound.

Location was underneath the buttress of a dead tree. Depth of burrow was about 05 m (f).

Factors considered in caging giant rat

In the study areas, burrows were mostly located under objects such as roots of trees, piles of dead trees, and stones. The role of these objects might be to provide insulation against heat. This is suggested by the fact that the areas of high concentration of giant rat were well shaded. However, shade and associated coolness might not be the only factor influencing the distribution of burrows. Such factors as availability of food, soil structure, and possibly minerals might also have some influence on their distribution. This implies that information obtained may not be conclusive on the factors that influence the distribution of giant rat burrows.

Hill et al. (1955) found that the extensive burrow system of the naked mole rat *Heterocephalus glaber* in Kenya provided a low-temperature habitat. Kristenson (1959) also found in Sweden that temperature variations are considerably reduced inside burrows. The evidence from the present findings is that giant rats prefer cool environments in which to locate their burrows. This implies that as far as possible, when placed in captivity, they should be provided with a cool environment.

The dimensions of burrows and burrow patterns reported in this study suggested an optimum habitat space for giant rats in captivity. The various burrow patterns also suggested that giant rats are clean animals, frequently having separate chambers for food and feces. It is therefore implied that these components should be given some considerations in managing the giant rats in captivity and to ensure that any husbandry technique that may be designed to raise them in captivity places an emphasis on regular cleaning of their environment.

It was possible for giant rats in the University of Ibadan halls of residence to raid kitchen stores and refuse areas, and these were the main sources of their food. Food remains inside the nesting chambers of rats at the zoological garden reflected what food was available in the vicinity. The study has provided information on the type of food on which giant rat fed on in Ibadan district and the relative importance of the food items. The list of food items resulting from this study, while it is not exhaustive on the species since similar studies were not made in other ecological zones outside Ibadan district, has provided information on possible kinds of food that could be provided for rats in captivity.

Fruit of the oil palm seemed to be the most important food item of giant rat in nature. The remains of this fruit were found in all the burrows excavated, and fragments of the kernels were also present in all the stomachs examined. It also formed a high proportion of the total diet consumed by each rat. The importance of this food can be readily appreciated when one examines its nutritive value. Oyenuga (1968) reported that except for methionine, the protein of palm kernel meal contains the essential amino acids and it is therefore of some value to the nonruminant animals. Palm kernel meal contains 20% protein. He also found the fatty acids dominating in palm oil (from the pericarp) to be oleic and palmitic and those in palm kernels (endosperm) to be lauric, myristic, and oleic. Thus the palm fruits and seeds are very rich in fatty oils that are readily hydrolyzed to their constituent fatty acids, in addition to their high energy density. Ether extract of palm kernel concentrate was 8.32% and 9.0% nitrogen-free extract. Thus, the fruit of the oil palm and its kernel alone provides almost a balanced diet for the giant rat.

However, all other components of food materials observed in the burrows also contained the essential amino acids (Oyenuga, 1968). Therefore the reason why palm fruits and kernels are specially preferred by giant rats is not clear. Further research is necessary to answer this question.

Factors which influenced the design of cages for the giant rat

Cage designs for the domestication of the African giant rat were based on the earlier field observations on their behavior and the characteristics of their habitats. Since giant rat burrows were located in cool places, temperature and ventilation were considered to be important factors that could influence the design of cages for rats. The burrow pattern of rats also showed that they had sanitary and food areas as well as exit holes. Rats are also nocturnal and monogamous. These facts suggest that any successful cage designs may have to simulate as much as possible the burrow pattern and also satisfy the nocturnal and monogamous behavior of rats. The area of nesting chambers recorded provided guidelines for the minimum space requirement of each rat in captivity. While satisfying the behavior of rats, cages must also be built to keep predators out. Cobras (*Naja* spp.) are the main predators of giant rats in nature (Rosevear, 1969).

A description of three kinds of cages designed to rear the African giant rat in captivity

The following is a description of three different kinds of cages that were designed to rate the African giant rat in captivity at different stages of their development.

Experimental feeding cage

This cage was designed to house one rat from weaning age to mature weight. Their growth rates on a particular diet, the amount of food consumed daily, and the digestibility of these diets were also observed in the cage. The components of the cages (Fig. 23) are as follows:

A collecting steel tray as the bottom of the cage for collecting urine. The tray was separated from the rest of the cage.

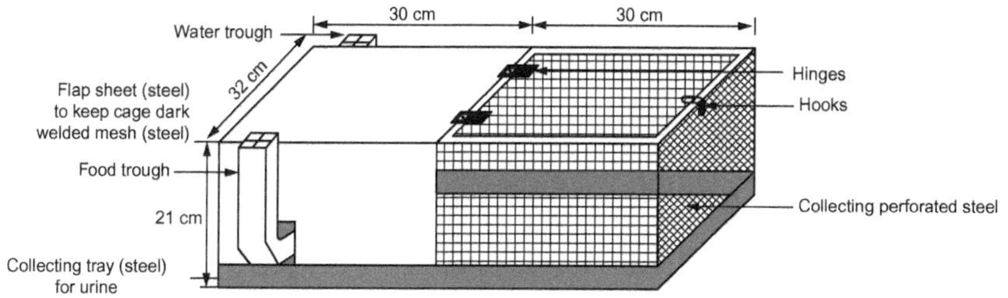

Fig. 23 An experimental feeding cage for the African giant rat.

A perforated steel for collecting feces but that allowed urine to filter through into the colleting tray later. The perforated steel sheet was built to slide freely in and out from one side at the bottom of the cage, so as to allow the removal of feces.

The main body of the cage consisted of a rectangular box of 1.8-cm screen wire and 2.3-mm thick mesh and had a floor area of 32cm × 60cm and a height of 21 cm. The screen wire could not be bitten through by rats, and the mesh was small enough to keep predators out. Half of the box was kept dark by means of a metal cap in order to satisfy the nocturnal behavior of the rats and to provide it with a nesting area. At this end, food and water troughs were also fixed to the wall of the cage each by means of two hooks, one at the top and the other at the bottom. It was necessary to keep this particular half of the cage dark because in nature, the dark sleeping chambers where rats normally rested during the day was usually the place where food was stored. If food was placed far from the dark end of the cage, rats had a habit of transferring food in their pouched jaws to the dark end where they rested during the day. In this way, particles of food could be scattered all over the cage and mixed with feces, thus making it impossible to make an accurate determination of the amount of food actually consumed. The other half of the cage with screen wire walls provided ventilation.

The "rehabilitation" cage

The "rehabilitation" cage was designed to house rats temporarily for about 1 month in captivity when they were newly trapped so that they became accustomed to living in cages. The components of the cage (Figs. 23 and 24) consisted of four equal compartments that were designed to hold a maximum of four rats. Each compartment had a solid wooden wall nesting box and a playroom with the same floor area of 0.45 m × 0.30 m. Aluminum trays were placed on the floors to collect urine and feces. The nesting box opened into the playroom by means of an entrance 12 cm × 15 cm.

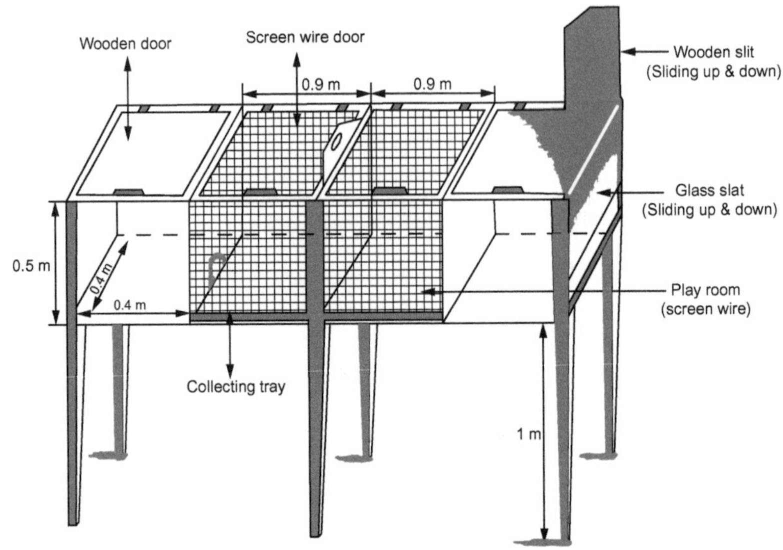

Fig. 24 A "rehabilitation" cage with the wire—gauze door opening backwards.

Each compartment had a door made of screen wire. Behind the door were two plywood partitions 9.6 cm × 0.39 m × 0.45 m one by the nesting box and the other by the playroom. The plywood partitions could be slid up and down in 12-mm-wide grooves. When both were slid up, the rats could be viewed in the compartment through the screen wire door.

The playroom was cleaned, first by opening the screen wire door backward, rising up the plywood partitions of the playroom to about one quarter the height of the cage, and by means of an iron rod, gently driving the rat through the door to the nesting chamber. The entrance of the nesting chamber was then closed by means of an aluminum door that slid up and down on the entrance. The collecting tray was then cleaned and replaced, the entrance reopened, the plywood partition of the playroom slid down to rest on the tray, and the rat driven back into the playroom. The process was then repeated for the nesting chamber.

The breeding cage

This was a "duplex" cage with two wooden nesting boxes at both ends of a rectangular playroom. The components are shown on Fig. 25. The playroom was divided into two halves by means of a plywood. The playroom was made of chicken wire 1.8-m mesh to keep snakes out of the cage and reinforced by a strong fence wire-netting 7.2-cm mesh and 3-mm thick to prevent rats from biting their ways through the cage. The playroom and nesting box had floor areas of 0.4 m × 0.4 m and 0.9 m, respectively.

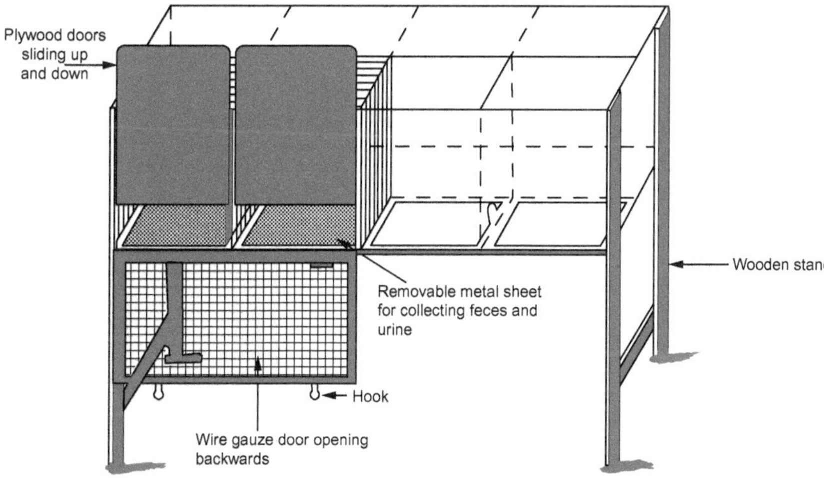

Fig. 25 A breeding cage.

At the back of each nesting box were plywood and glass dividers, which slide up and down on an aluminum collecting tray. When the plywood was raised up, rats were observed in the nesting box through the glass. The nesting box opened into the playroom by means of an entrance 12 cm × 15 cm. The nesting box was cleaned first by "persuading" the rats to move from the nesting box into the playroom and closing the entrance. The plywood and glass partition at the back of the nesting box was then raised up, the underling collecting tray removed and cleaned. The playroom was also cleaned by allowing the rats to go back into the nesting chamber, closing the entrance and sliding out the collecting tray. Each half of the breeding cage was designed to hold a breeding pair of rats or a nursing female with its young up to weaning age.

Management of rats in cages

The behavior and burrow patterns of giant rats in their natural habitats showed that they are clean animals. Even when the same nesting chambers of a burrow served functions such as food storage, for sleeping and defecating, different sections of the chamber were used for these purposes. Daily cleaning of cages was therefore considered important to the health of the animals. All collecting trays for wire and feces were washed daily. Food and water were served in aluminum containers and placed in the playrooms. These containers were also cleaned daily. Newly trapped rats and the breeding colony were fed ad libitum on balanced commercial rabbit diet according to UFAW (1971) and supplemented daily with oil palm fruits and kernels that seemed to be their most favored food item.

Each compartment had a door made of a screen wire. Behind the door were two plywood partitions 9.6 mm × 0.39 m one by the nesting box and the other by the playroom. The plywood partitions can be slid up and down in 12 mm-wide grooves. When both were slid up, the rat could be viewed in the compartment through the screen wire door.

Reproductive performance

In order to assess the merit and demerit of the African giant rat as a producer of meat, observations were made on its reproductive performance. Factors of investigation included the following:

a. Age at sexual maturity
b. Estrus cycle
c. Gestation period
d. Age at which young rats are weaned
e. Birth records, for example, dates of birth to find out if births are seasonal or not, litter size of each litter, individual birth weights and average weight of each litter, standard length from nose to anus, and total length (from snout to tip of the tail)

UFAW (1971) found that there was a marked effect of litter size on birth weights and number of birth weights on weaning weights of rabbits. They also found an inverse relationship between size and individual birth weights. Larger birth weights were also found to give rise to heavier weaning weights and vice versa. Female giant rats attain puberty at about 20–23 weeks. They exhibited an estrus cycle of about 4 days. The gestation period is about 28 days, and young rats are weaned at 21–26 days. Male–female sex ratio is 1.2:1.

The highest number of litters observed per female was five in nine months. It is possible that a female can reproduce six times in 1 year. With an average litter size of four, a female can therefore be expected to produce about 24 young per year. Apart from cannibalism, a phenomenon that did not appear to follow any fixed pattern, juvenile mortality in giant rats was found to be negligible. The giant rat reproduced throughout the year and did not show any seasonal peak in breeding. This fact, with a relatively short estrus cycle of about 4 days, can be an advantage in gestation, since meat production from this rodent can be ensured throughout the year.

Behavior of rats in captivity

Wild rats are ferocious, but their offsprings if handled regularly when they are young and usually before their eyes opened became tamed and showed no tendency to bite. However, there was still a tendency to escape from the cage when it was not properly locked. The rats of the first generation constantly chewed the entrance of their nesting box. Among the second and third generations, some components of fighting behavior such as threat posturing, ran chattering, leaping, and biting have been totally lost. Some aggression occurred when rats were paired for the first time, but this was mild and relatively harmless and resembled the playful bottling of immature rats. Wildness and savageness toward humans remained only with nursing mothers of the second generation, but this was completely lost in the third and fourth generations. The third and fourth generations of rats were very docile and, when placed on the floor of the normal house, showed no tendency to escape. Incidence of "cage chewing," which was typical of wild rats, however persisted.

Among characters that seemed instinctive and that persisted to the fourth generation was their preference for palm fruits *Elaeis guineensis*. When a mixture of food items selected from one that formed the rats' diets in the wild was presented to them, they greedily consumed all the palm fruits first before eating others. This happened even in cases when the young rats had no diet of palm fruits prior to the experiment. Figs. 26 and 27 show docile giant rats of the third generation being handled freely.

Care of progeny

Rats showed considerable care for their young. Nursing mothers might permit suckling for about 3 months after the weaning age. Mothers spent a considerable period licking their young especially during their first 2 weeks of birth. Fig. 28 shows a nursing mother with its young in the breeding cage.

Eating and grooming

When eating, giant rats sat on their haunches and held the food with their forefeet. When fed communally, they rushed at the food, in the playroom, stored it quickly in their cheek pouches, and deposited it at different areas of the nesting box where it was eaten slowly. Several runs were made by one rat before the serving bow was empty. After eating, it licked its forefeet until they were wet, and they then rubbed on its snout. Giant rats are easily

Fig. 26 Wild male and female giant rats exhibiting agonistic behavior in a "boxing" position. The author (first right) (Photo 1973).

Fig. 27 The author, (right) inspecting a docile giant rat after its rehabilitation. Extreme left was Professor Laurence Roche (late): Author's PhD Supervisor (Photo 1973).

tamed and readily adapted to new diets. A low level of aggression was noticed among rats of the first generation and almost a complete absence of fighting behavior and aggression in the subsequent generations even when they were paired for the first time. This suggests that giant rats can be managed to become social animals.

Fig. 28 The African giant rat nursing its young in a breeding cage.

Sleeping

Individual rat usually slept either in a curled position with its head placed against its abdomen or on its back or flank like humans (Fig. 29). When it awoke, it yawned and stretched by extending its forelimbs and hind limbs, dragging itself forward on the belly. When it slept, its hair often appeared fluffy, and it breathed more slowly than when awake and active.

Fig. 29 Sleeping position of the African giant rat.

Pathological conditions of rats in captivity

Observations were made on the pathological conditions of 22 rats that died in captivity within 3 months of study. Several animals had more than one pathological condition. The following were the groups of observations made according to management problems and infectious processes:

(i) *Management problems*

There were two problems encountered during the domestication study that could be important to wild animal husbandry. First, giant rats being burrowing rodents preferred a cool environment in nature, with a low tolerance to heat. This problem was however overcome when the breeding colony was transferred to a well-ventilated and shaded animal house (Ajayi, 1974). Secondly, cannibalism had persisted throughout the period of study, even when rats were fed on balanced commercial rations of rabbit pellets that UFAW (1971) recommended for the management of rodents in captivity and supplemented with palm fruits and leaves of *Centrosema pubescens,* a proteinous leguminous plant, to provide additional vitamins, minerals, and protein. It therefore appeared that cannibalism was not as a result of nutritional deficiency. UFAW (1971) has however found cannibalism to be an important management problem among laboratory rodents in general even when they were raised under seemingly ideal conditions. Condition of Overeating and "Tympany": This problem was also diagnosed in four young rats that were introduced to new food rations with high grain constituents. Animals had large quantities of dry or moist concentrate of the food ration in their distended stomachs. Associated lesions were distended abdomen and acute pulmonary congestion.

(ii) *Heat exhaustion*

Excessive heat was believed to be the cause of acute circulatory failure that was observed in four rats, which died suddenly between the dry hot months of January and March of the study period when the ambient temperatures were above 32°C. The animals had severely congested and edematous lungs and congested abdominal viscera, especially the liver. This problem was eliminated when rats were removed from a temporary enclosure and housed in a larger and more ventilated animal house that was well shaded by the surrounding *Eucalyptus* and banana trees.

(iii) *Fighting*

Severe fight wounds were found in the body of four rats. The wounds were freshly inflicted in three rats, and on the fourth, wounds have been infected before it died. Fighting occurred when newly captured wild rats were paired in the breeding cage for the first time or when wild rats exhibited escape reactions in the "rehabilitation" cage, thereby bruising their head and other parts of the body.

(iv) *Urinary cystolithiasis*

The urinary bladder in a male newly captured rat that died of heat exhaustion contained numerous small, light-yellow, friable calculi that extended a few centimeters into the urethra without causing complete obstruction. There was no evidence of urine retention or hydronephrosis. These renal calculi might have developed in the wild before the rat was subjected to captivity. This is because it died after only about 1 week in captivity during which water was provided ad libitum.

(v) *Infectious processes*

The infectious processes were due to parasites and bacteria. The parasitic infections were taeniasis in wild rats (observed in three animals) and capillariasis (observed in one animal). The tape worms in rats with taeniasis were *Hymenolepis diminuta* and *Hynenolepsis nana*. The liver of the wild rat that died of capillariasis had its liver grossly enlarged and contained numerous, irregular ova of Capillaria with grayish-white nodular appearance. Bacterial infections were characterized by fibrinous purulent peritonitis in two rats and purulent dermatitis extending to the underlying muscles in two other animals. Additional lesions were perirenal abscess extending to the cortex of one kidney in one rat and postpartum purulent metritis with metastasis to the peritoneum and meninges in one animal. *Escherichia coli* was isolated from the uterus of the latter. The earlier infectious processes in captured rats have provided an indication of the disease problems to be expected among wild rats in the course of domestication. However, the problems were solved by the administration of drugs to infected rats. That was why subsequent generations of rats were remarkably free of parasites and bacterial infections.

CHAPTER 17

Ex situ conservation of wildlife: Domestication of the grass cutter (cane rat) *Thryonomys swinderianus*

Introduction

The cane rat, otherwise known as the grass cutter *Thyronomys swinderianus*, is a rodent that is closely related to the porcupine, capybara, and guinea pig. It is found in forests and savanna areas of most parts of Africa, South of the Sahara. It is entirely herbivorous, and its favorite food is the succulent nourishing internode of savanna grasses, particularly, the elephant grass *Pennisetum purpureum*. The grass cutter shelters in hollows of trees but does not burrow. Forest clearance for agriculture of the savannah and tropical forests has expanded its habitats into secondary savannas, crop areas, and sugarcane plantations. Although they can thrive entirely on grass, they also show preference to tubers of sweet potato and cassava. Cane rats have therefore been incriminated as pests of farm crops, including rice.

The process of establishment of a breeding colony of grass cutter

Trapping of wild rats

One of the easiest and most practical ways of trapping grass cutter to establish a breeding colony is by the use of nets on sugarcane plantations. Grass cutter populations are relatively high on sugarcane plantations, particularly in the savanna regions of Africa. Local communities have perfected a method of skillful placement of nets in a particular direction when the farm is being burnt for harvests of sugarcane. Grass cutters that run into the nets are trapped by a careful but swift manipulation of the nets, sometimes assisted by dogs. Although this method is fast, it nevertheless has its own disadvantage, since it can only be used during the sugarcane harvesting season. However, local communities have used this method elsewhere where cane rats are found in large numbers.

Wildlife Conservation in Africa
https://doi.org/10.1016/B978-0-12-816962-9.00017-X

The second method is by the use of a standard "Havahart trap" size 4. This is a perforated rectangular box with openings at both ends and the trigger system at the middle. The bait, which is a potato or cassava tuber, is placed firmly on the flat metal plate of the trigger inside the box. Grass cutters are trapped when the trigger is released closing the iron flaps at both ends of the trap. Although grass cutters favor potato and cassava tubers, the most effective bait for this animal is human urine on a lump of soil. Grass cutter's preference for human urine is so strong that places where urine is deposited on farmlands are always intensively licked and scattered. Although this method of trapping ensures the safety of captured animals, it is nevertheless a very slow process to establish a parent breeding stock.

Grass cutters can also be trapped by digging holes of about 1 m deep on their "high way" and covered with leaves and grass. When recovering live wild animals from such holes, they should be held by the hip and not by the tail to prevent escape.

Hunters' techniques: Grass cutters move in colonies or parties with the smallest in front. When adult females and males are shot during the breeding season in West Africa (August/September), the animals lose their orientation and scatter. Hunters normally look around to collect the young animals that are in confusion after the adults have been shot.

Note that to start a new breeding colony of grass cutters, preference must be given to young animals (less than 3 months old) and preferably should not weigh more than 1 kg live weight. It is desirable to start domesticating with young animals whose wild instinct has not been fully developed. At a young age, wild grass cutters are easy to manipulate.

Transportation of wild grass cutters

Wild rats are nervous, aggressive, and vicious and therefore should be handled with hand gloves. Handling by the tail should be avoided since the tail is easily removed by the slightest force. Handling of wild grass cutters should be by the hind legs and hips. When rats are tamed and have become docile, handling can be done safely by the tail. Wild grass cutters are usually transported from their locations of capture to domestication destinations, through long distances in wooden cages with wire gauze sidewalls for ventilation. The top has an "open/lock" catcher mechanism, to let rats in and out of the cage (Fig. 30).

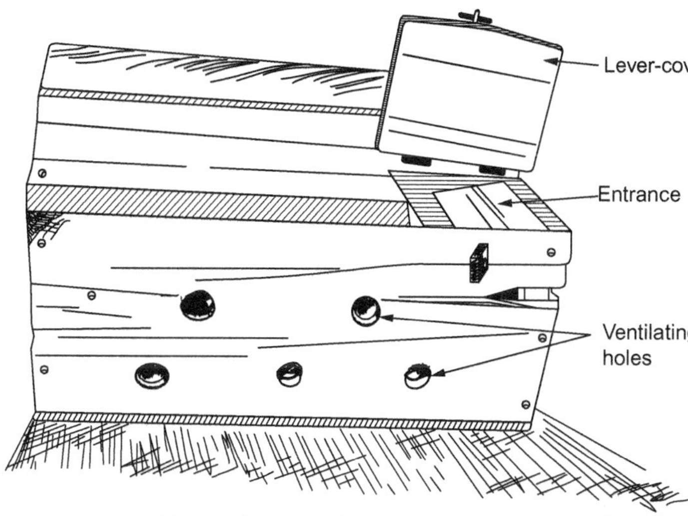

Fig. 30 Grass cutter transportation cage—University of Ibadan model.

Cage designs

In order to design cages for the grass cutter in captivity, its habitat requirements and behavior were considered and simulated as much as possible. Grass cutter is a nocturnal rodent, with most of its feeding activities taking place at night. This was an important factor in designing its enclosures (Fig. 30). Temperature and relative humidity were also found to be important parameters for keeping them successfully as farm animals. For example, the relative humidity should not be lower than 45% and the temperature not less than the ambient temperature of about 20–26°C. A relative humidity of 50%–60% with the temperature slightly higher than ambient temperature is however ideal for the grass cutter. Using the earlier factors and information on its habitat requirements, a farmhouse (Fig. 30) and the following cages have been designed to rear cane rats successfully in captivity (Ajayi, 1976; Lameed, 2008).

Management systems of grass cutters (cane rats) in captivity

There are three major management systems of grass cutters (cane rats) reported in the succeeding text: (A), (B), and (C).

Semifree range management system

The first is a semifree range management system whereby a simple rectangular house is provided (Fig. 31). A typical farmhouse for a semiintensive and semifree range management of grass cutter is a simple, rectangular bungalow about 11 m × 15 m (Fig. 31). It must be provided with an iron roof, supported by iron poles and wire netting walls to allow abundant ventilation for the animals. The floor and foundation are cement concrete, and there is a wall around the house, which is about 1½ m high, upon which the wire netting rests. Fig. 31 provides a simple illustration of a grass cutter farmhouse. Several "hutches" are provided on the concrete floor of the grass cutter house. Details of the structure of "hutches" are shown in Fig. 31. Grass cutters can move freely in and out of their respective cages for feeding and drinking and later retreat to their respective hutches for sleeping and breeding. Details of this system are described in Figs. 32–34.

Housing for a semifree range system

The semifree range is a low-input system for small-scale farmers; it is designed to minimize cost without losing the potential of the grass cutter as

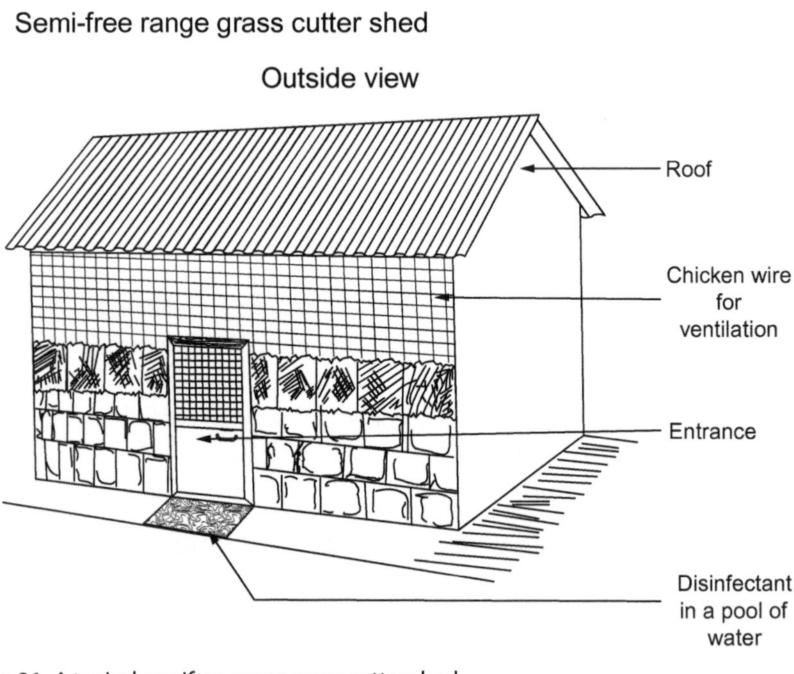

Fig. 31 A typical semifree range grass cutter shed.

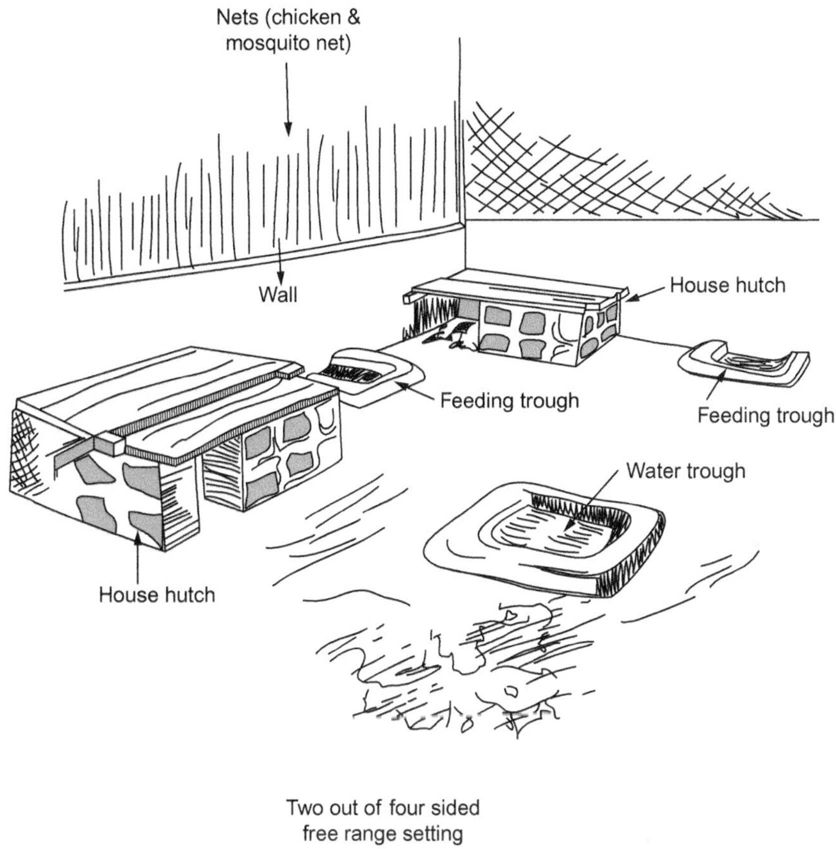

Fig. 32 Floor diagram (inside view) of a semifree range grass cutter house with locations of "hutches."

Fig. 33 Domesticated, docile grass cutter being handled freely by Professor G.A. Lameed of the Department of Wildlife and Ecotourism, University of Ibadan, Nigeria.

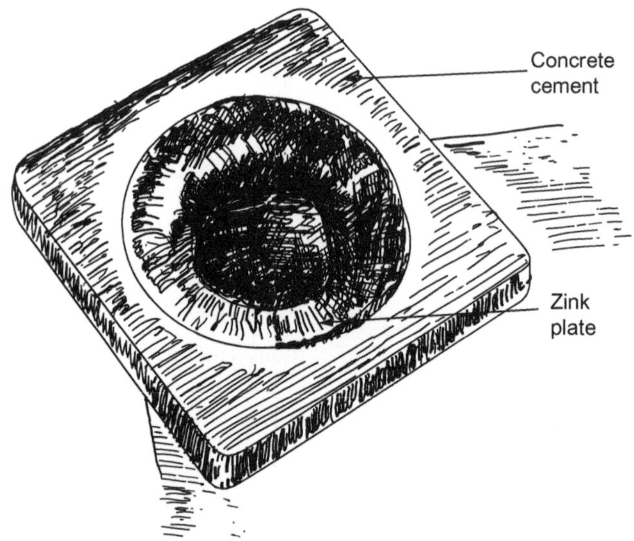

Fig. 34 Concrete feeder for free range grass cutter on a cement floor.

a producer of meat. The simple rectangular house has sidewalls made of chicken wire, resting upon a meter high concrete wall. The roof is made of zinc sheets, and it is usually slanted, about 1½m beyond the walls, to prevent rain from splashing it. It is important to provide a gutter around the house with antiseptic and insecticide solutions in it to prevent ants and human infections from getting in.

Husbandry in semifree range systems

Grass cutters from their wild natural habitat are reared on a bare concrete floor. The floor consists of "hutches" or cane rat "units" with cement block walls as shown in Fig. 32. The top of each hutch is covered with planks to prevent escape of rats. The dimensions of each hutch are length 1½m, width 1.0m, and height 05m. The main feature of a "hutch" or "unit" is shown in Fig. 32.

Fig. 34 shows a concrete feeder for solid food for grass cutters (cane rat).

Intensive management system

The second system is intensive and requires more capital outlay than the semifree range system. It consists of a battery of concrete cabins constructed on a bare floor of a typical grass cutter house. Each cabin is subdivided into a dark "bedroom" and a lit "playroom." This is to satisfy their nocturnal and diurnal behavior. Details of the structures and operation procedures are provided in Figs. 33–35.

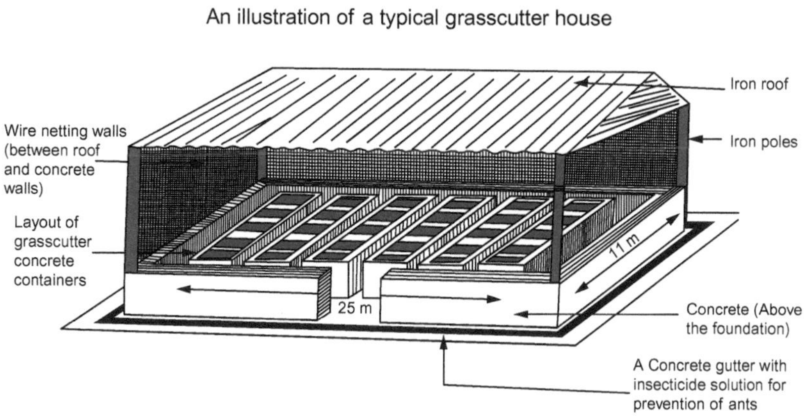

Fig. 35 An illustration of floor plan of a grass cutter farmhouse. It contains several concrete cabins that are again subdivided into a dark "bedroom" and a lit "playroom."

Illustration of structures for an intensive management of grass cutter

Fig. 35 is an illustration of a floor plan of a grass cutter farmhouse. Each grass cutter unit consists of a darkroom, for sleeping and mating to satisfy their secretive, nocturnal behavior, and an open or "playroom" where food and water are provided. The dimensions of each unit are length 1½ m × width 1.0 m × height 0.05 m.

Caging

The house contains a layout of a battery of concrete cages on a bare concrete floor. As it is with the domestication of the African giant rat, published information on wild animal husbandry has shown that the management of rodents in captivity is usually based on caging. The account of past efforts in managing rodents in captivity shows that grass cutters can also adjust to the abnormal conditions of captivity and that they thrived well in concrete cages. Fig. 36 is an illustration of floor plan of a grass cutter farmhouse. It consists of a battery of concrete cabins called "grass cutter units" or hutches that are subdivided into a "dark bedroom" and a lit "playroom" (Fig. 36).

Intensive management of grass cutter in a three-tier cage system

Another intensive management system is by utilizing a similar grass cutter house as in Fig. 37 for an arrangement of wooden cages in three tiers. Details of the arrangement and structures are provided in Figs. 38–40.

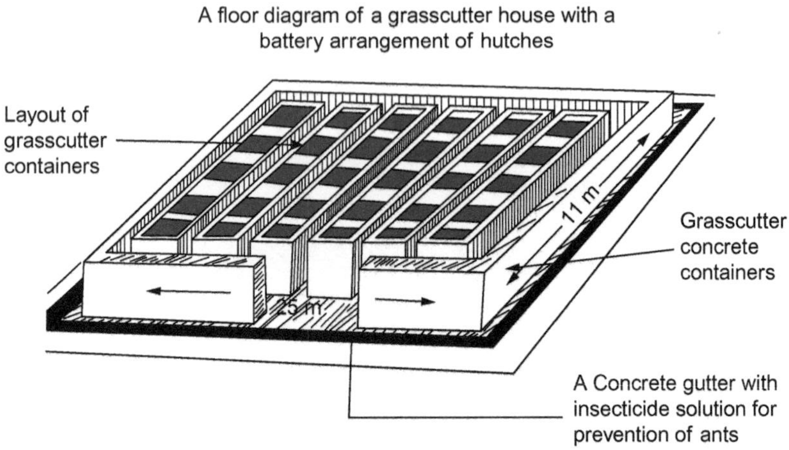

Fig. 36 A floor diagram of a grass cutter house with a battery arrangement of hutches.

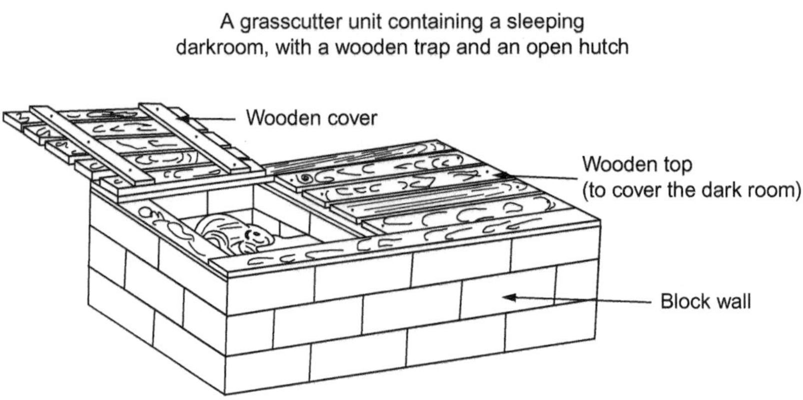

Fig. 37 A grass cutter "unit" containing a sleeping darkroom with a wooden top and an open wooden cover.

In a typical grass cutter house described in Figs. 38 and 39, there is a battery arrangement of a three-tier cage consisting of three layers of wooden cages. Each cage consists of "two grass cutter units" as follows:

A self-contained unit with a darkroom and a lit open parlor where grass cutters are paired one male to four females for breeding purposes. Fig. 38 shows the room and parlor displayed on the right hand of the cage. To the left is a "restraint" room where pregnant females are kept before the delivery of their young. The dimensions of the three-tier cage are shown on Figs. 38 and 39.

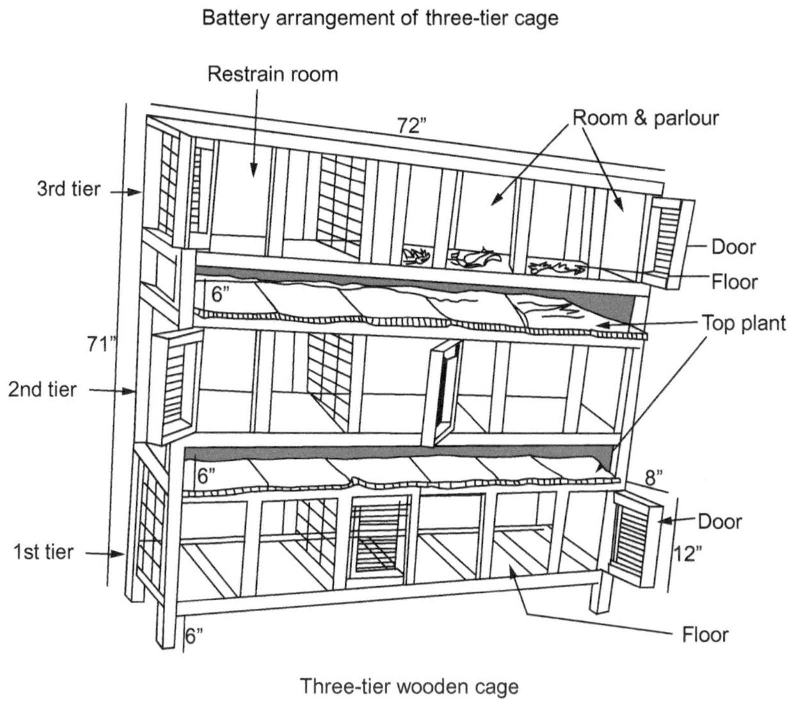

Fig. 38 Battery arrangement of a three-tier wooden cage for intensive management of grass cutter (the "restraint" room is designed to hold pregnant/nursing female with its young up to weaning age).

On the floor of each cage is a perforated metal sheet that permits urine to flow to a metal collecting tray that lies at the bottom of each tier of cages (Fig. 40).

Grass-cutter behavior and husbandry

Rehabilitation of a wild colony in captivity

Wild rats are usually tamed to docility in about 28–30 days. It is important that at the transitional period from the wild status to captivity, grass cutters are carefully manipulated to ensure a gradual elimination of shyness, stress, and aggression.

The grass cutter house and the enclosures should be shaded by vegetation to provide a dark environment because they are nocturnal animals.

Since they are animals of forage, it is important to ensure a constant supply of grass, particularly their most favored food item, elephant grass *Pennisetum purpureum* (Fig. 41).

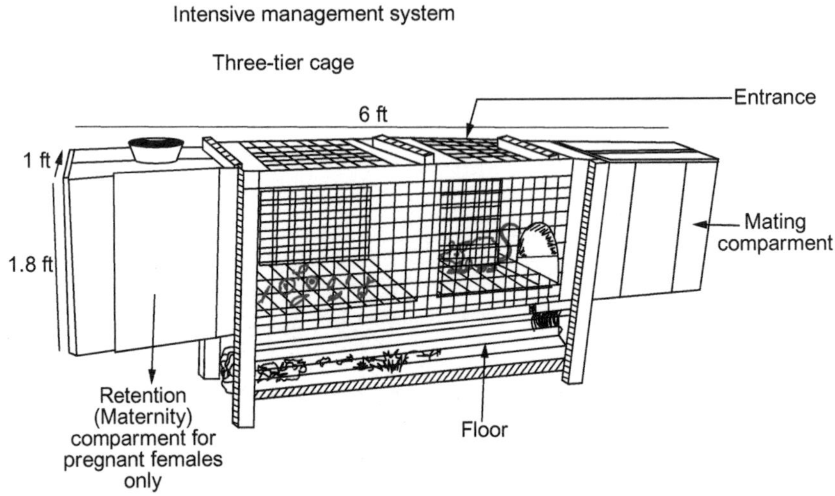

Fig. 39 The three-tier cage system with details of composition of one tier showing the mating component and the retention (maternity) compartment for pregnant females.

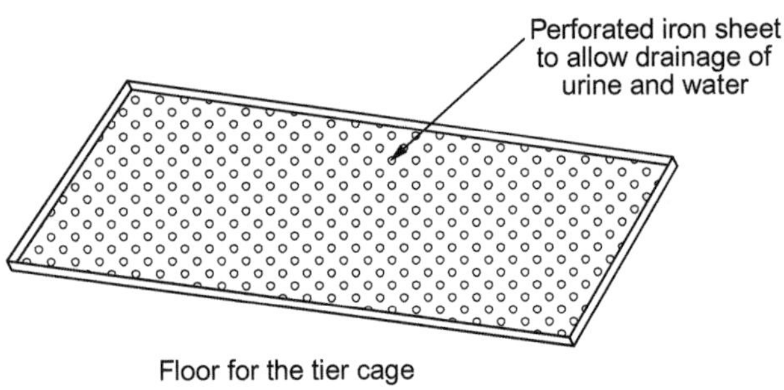

Fig. 40 Perforated iron sheet that forms the floor of each tier of the cage that allows urine and water to filter to a collecting tray underneath.

There should be a constant supply of water and sugar solution, known as "sweet water." This is important for rearing grass cutters in captivity.

Salty food, that is, food with sodium chloride, should also be supplied daily.

In order to avoid shock, wild grass cutters should be gradually exposed to "movements" and sounds. The farmhouse attendant must wear only one particular dress during the first 1–30-day transitional period. If there is an abrupt change of these features, the wild stock may react negatively to them.

After about 21 days in captivity, the animal attendant may gradually reduce the darkness created around the grass cutter "cabins."

Fig. 41 Elephant grass *Pennisetum purpureum* favorite food of the grass cutter.

Coccidiosis is usually their initial disease problem. Therefore all animals should be vaccinated from their day 1 in captivity as follows:

One cubic centimeter of colchicine/8ND for each animal by intramuscular injection: Injection must be applied to the muscle between the root of the tail and hind legs. The space here is "V" shaped. It is important to note that this treatment should be done immediately to the wild rats that are subjected to captivity and should be done only once in its lifetime.

Management and adaptation to captivity

The first 7 days within the enclosure: During the first day of release, grass cutters usually hid under a pile of grass at all corners of the enclosure. They did not eat any of the food items provided. The little movement observed was usually for the assessment of their environment presumably to ensure their safety. From the second day, newly captured rats started to eat familiar food items such as grass and tuber. From about 8.00–12.00 h in captivity, they ventured out of their confinement to eat. This was the general pattern of behavior for the first 6 days, but on the seventh day, they ate more voraciously, and a larger portion of the elephant grass was consumed. At this stage, the grass gutter was presumed to have overcome the shock of the "abnormal environment of captivity."

Adaptation to captivity (day 8–14): From the eighth to ninth day, captive rats began to exhibit a gregarious behavior as they started to congregate in groups (even if they have been captured from different localities). They ate well during the day, but their feeding period was predominantly at night.

From the 10th to 14th day, extensive movements around their enclosures took place particularly at night.

Adaptation of captivity (day 15 –28): Wild grass cutters have become acclimatized and used to humans to the extent that they no more reacted to human presence. They came out of the darkroom to eat during the day. By the 28th day, grass cutters were so accustomed to their confinement that they began to move freely in the "darkroom." Although they become relatively tamed after about 30 days, the general behavior still showed that grass cutters are slow to adapt to humans.

Feeding

Although grass cutters can feed exclusively on fresh elephant grass and tubers such as potatoes and cassava, they nevertheless adapted quickly to supplementary feeds such as groundnut cake and maize mill and to commercial animal feeds, such as poultry and pig rations. Other supplementary feeds include maize chaff and brewer's spent grains. Its general food consumption rate on grains, grass, and other foodstuffs is shown on Tables 46 and 47 (Ajayi and Tewe, 1980).

Productivity and economic considerations of grass cutter farming

According to Ajayi (1981), grass cutter reproduces twice a year with a maximum litter size of 6–8 young. It attains physical maturity of about 5.8 kg in 8 months. It is possible for males to attain a weight of 10 kg in 14 months. It has a killing out percentage of 63.8, and it is a very rich source of protein and minerals. In practical terms, given this reproductive and growth

Table 46 Average daily consumption of feeds by cane rats for phase 4 ($n=6$).

	Food consumption (g/day)	
Food item	**Fresh basis**	**Dry matter basis**
Elephant grass	$387.0^{a}+4.4$	$135.5^{a}+6.1$
Maize mill	$11.3^{b}+1.3$	$9.6^{b}+1.4$
Commercial poultry ration	$3.5^{b}+0.9$	$3.2^{b}+2.4$
Commercial pig ration	$1.8^{b}+0.3$	$1.6^{b}+0.4$
Groundnut meal	$1.1^{b}+1.1$	$0.99^{b}+0.6$
Palm kernel cake	$1.3^{b}+0.06$	$1.2^{b}+0.03$

Means with common superscripts within columns are not significantly different from each other. Others differ at a probability level of 0.05 or less.

Table 47 Average daily consumption of feeds by grass cutters and standard error of mean of grass cutters ($n=6$) (Ajayi and Tewe, 1980).

Food item	Food consumption (g/day)	
	Fresh basis	Dry matter basis
Elephant grass	$401.4^{a}+10.2$	$140.4^{a}+10.2$
Maize grain	$14.8^{c}+1.4$	$13.8^{c}+6.8$
Rice	$19.2^{c}+4.8$	$16.3^{d}+0.9$
Guinea corn	$42.7^{b}+0.2$	$38^{b}+2.3$
Groundnut	$19.1^{c}+2.3$	$17.67^{c}+6.9$

Means with common superscripts within columns are not significantly different from each other. Others differ at a probability level of 0.05 or less.

potential, a female can produce 14 rats in 1 year, that is, a total of 94.4 kg live weight or 62.9 kg dressed carcass of meat. This means that in about 20 months, a population of 10 females and 2 males of grass cutters can produce about 620.9 kg of meat.

Farming grass cutters can be a profitable venture only if the capital outlay and the running costs are made as negligible as possible, so as to break even at the third or fourth year of operation. There is an obvious economic advantage in its feeding habits and food preferences. As a monogastric herbivore, it can thrive entirely on fresh grass and therefore needs not compete with man for its food. The protein requirement is about 16% for adults and 18% for young rats.

The housing should be made as simple as possible but should be snake- and dog-proof. Daily husbandry when the animals have been accustomed to captivity with the night temperature and relative humidity will ensure an establishment of a healthy, viable colony.

Feeding of grass cutter colony

Grass cutters can be fed daily on a combination of all the items in the succeeding text:

1. Elephant grass
2. Maize
3. Brewer waste
4. Palm kernel waste
5. Table salt
6. Calcium

Charcoal is usually provided (mixed with water and items 2–6 above) to prevent diarrhea.

Feeding time table

The following is the suggested routine/daily feeding time table for a grass cutter colony:

At 9.00–9.30 a.m., supply only fresh elephant grass (where possible) and water.

10.00–1.00 a.m., this is the time for supplementary feeding with food items 2–6 above.

At 3.00–4.00 p.m., feed with elephant grass again, with fruits such as pawpaw and banana.

At 6.00–7.00 p.m., a mixture of supplementary feed items 2–6 with elephant grass.

In addition to the earlier time-specific feeding program, fresh elephant grass and, where available, potato tubers and "sweet water" should be placed daily in the open room where grass cutters feed ad libitum. Grass cutters are usually shy of humans at an early stage of captivity. It is therefore important that a lot of fresh grass is provided at each corner of the cage not only for feeding but also for hiding.

PART 4

Curriculum development and evolution capacity building for protected area management in Africa

CHAPTER 18

Curriculum development and evolution of capacity building for protected area management in Africa: The experience of University of Ibadan, Nigeria

Introduction

Since this book is meant primarily for students of wildlife conservation in tertiary institutions, it should be noted that the subject of wildlife cannot be taught in isolation from the other components of natural resource conservation that are also major components of protected areas. Wildlife, fisheries, and forestry are therefore important components of "protected areas" in Africa. In order to provide a broad base for students of conservation, wildlife management cannot be taught in isolation from other sectors of natural resources, particularly from 100 to 300 levels for undergraduate courses. Hence, the curricula presented here provide a comprehensive picture of major components of protected areas for conservation purposes in Africa.

Establishment of the Pioneer Department of Forestry, University of Ibadan, 1963

Capacity building is fundamental to effective conservation and restoration of Africa's depleted natural resources and biodiversity. Similarly, the quality of manpower produced and the overall impact on the environment depend to a large extent on the relevance and viability of the curricula upon which they are trained.

That was why Food and Agricultural Organization (FAO) took an important step to establish in 1963 (as far as it is known), the first African Department of Forestry at the University of Ibadan under a project plan of operation of United Nations/Food and Agriculture Organization (UN/FAO) "Special Fund Project 110." The objective was "to provide high-level

Wildlife Conservation in Africa
https://doi.org/10.1016/B978-0-12-816962-9.00018-1

education in forestry and protected area management and for the management of natural resources in Africa." After about half a century of evolution of capacity building in protected area management at the University of Ibadan, it is pertinent that the development of higher education in this sector be given a prominent attention. It is for this reason that the historical and consequent development of capacity building at the University of Ibadan in protected area management and, in particular, wildlife and fisheries management is presented in this chapter.

Perhaps, the most important contribution of the author of this book to the wildlife conservation in Africa was in the area of manpower and curriculum development. As far as it is known, he was the first African to obtain a PhD degree in wildlife management and indeed the first Nigerian to be appointed a lecturer in this field. Furthermore, he was the first African to be appointed a professor of wildlife management.

It is important to provide a brief account of his experience in founding the Department of Wildlife and Fisheries Management at the University of Ibadan and the development of various curricula. The study of wildlife in Ibadan owed much to the FAO, which in 1968/69 academic session, as part of its staff development program, commissioned a wildlife biologist, Dr. W.N. Holsworth, to introduce a wildlife conservation course into the curriculum of the Part 1 students only in the then Faculty of Agriculture, Forestry, and Veterinary Science. The course was titled "Agriculture and Climatology," and wildlife formed only about 50% of this three-unit course or less than 0.05% of the whole agriculture curriculum. It was in the same academic session that the author joined the University of Ibadan as a temporary assistant training fellow in the then Department of Forestry.

In 1970, FAO further introduced wildlife conservation into the curriculum of the then Forestry Department as part of a three-unit savanna ecology course, FOR 212. Wildlife issues constituted about 50% of this course or about 15 h of theoretical instruction time per year. The objective of the course according to the FAO expert was "to produce well-rounded foresters who would be versed in the management of forest products other than cellulose."

At the later part of the 1970/71 academic session, the author returned to Nigeria after completing his MSc degree at the University of Edinburgh, Scotland, and his stint at the College of African Wildlife Management, Mweka, Tanzania. He took over the responsibilities of the FAO expert and also started a wildlife conservation unit as a lecturer grade II in wildlife management. By the 1974/75 academic session, the activities of the wildlife unit had become prominent enough in the Department of Forestry to necessitate the restructuring of the curriculum to give wildlife and fisheries management more prominence. The name of the department was correspondingly changed to the "Department of Forest Resources Management" in the same session to reflect the pragmatic metamorphosis of its academic program from pure cellulose to all-embracing management of forest products notably, wildlife and fisheries.

How fisheries and wildlife management was introduced to the curriculum of forestry in 1975/76 academic year

It is important to state here that the background of the author in fisheries management as an employee of the defunct Northern Nigeria Ministry of Agriculture was brought to the attention of the then head of department late Professor Laurence Roche and to the Department of Forestry in 1975/76 session. A synopsis of the author's tenure and experience in fisheries is as follows:

> The author was appointed by the defunct Northern Nigerian Government as fisheries assistant grade II, that is, his first appointment on a salary of 16 Nigerian pounds on January 22, 1965. He was promoted over fisheries assistant grade I to the grade of assistant fisheries superintendent on a salary of 45 Nigerian pounds per month within 6 months of his first appointment. His elevation was due to his three principal passes in chemistry, zoology, and botany at a sitting in the (Cambridge Higher School Certificate) advanced level result in 1964. After the author's 3-year "in-service" degree program leading to the honors degree in zoology at Ahmadu Bello University (1965–68), he was elevated again to the post of fisheries officer on a salary of 60 Nigerian pounds per month in July 1968. His experience in fisheries management during this period included fish farming practices in Panyam fish farm in Plateau province (Plateau State). He was later posted to Funtua, Zamfara State, where he built Maska fish farm, the second fish farm in Northern Nigeria in 1965 by the Northern Nigerian Government. Prior to his posting to Maska fish farm, he undertook a survey of some parts of northwestern Nigeria to assess its potential for fish farming. Areas of particular interests included Talata Mafara (Zamfara State) and Lugu Dam in Wurno, Sokoto State, and of course Maska Dam, Zamfara State. These places were marked as potential areas for fish farming development in northwestern Nigeria. It was after the fisheries survey that

Maska fish farm was built, and that was shortly before he proceeded to Ahmadu Bello University for his degree course.

The author was at Ahmadu Bello University in Zaria from 1965 to 1968. During his vacation in July to September 1966, the Northern Nigerian Government seconded him to the UN/FAO Riverine Fisheries Survey Team to undertake the fisheries survey of River Niger from Lokoja upward to New Bussa, Shagunu, and Zaria Kalakala. The mandate of the fisheries team was the assessment of the fishes and the fisheries potential of the upper Niger for development in capture fisheries. Furthermore, during his 3-year degree course in zoology at Ahmadu Bello University and precisely in 1967, he had an intensive training in marine ecology and artisanal fisheries of Nigerian brackish and coastal waters at the University of Lagos by a special arrangement between the Department of Zoology, Ahmadu Bello University, and the Department of Biological Sciences, the University of Lagos, as an integral part of zoology program of Ahmadu Bello University. Thus, in a period of about 4 years, the author had an exposure to ex situ and in situ management of inland and marine fisheries. Hence, with the permission of his head of department in 1975/76 session, he brought his background and experience in fisheries management to the development of curriculum of forestry department at the University of Ibadan. Hence, he had a solid platform upon which to make a case to develop a unit of wildlife and fisheries management with animal concept in biodiversity as part of the curriculum of forestry at the University of Ibadan in 1975/76 academic year. This was the year when agriculture degrees and forestry in the Faculty of Agriculture and Forestry were translated from 3- to 4-year programs.

The wildlife and fisheries unit organized successfully an international symposium on "wildlife management in savanna woodland" in collaboration with the College of African Wildlife Management, Garoua, Cameroun, in 1975/76 session. The symposium focused on wildlife conservation problems and prospects in Africa and provided the foundation stone for the establishment of the Department of Wildlife and Fisheries Management at the University of Ibadan. It was at the time that the conference communiqué recommended the development of the wildlife unit into a full-fledged Department of Wildlife and Range Management in consonance with the heavy personnel requirements for the development of African wildlife and fisheries resources. It was the acceptance of this communiqué that led to the teaching and research work in wildlife conservation and fisheries management at the University of Ibadan.

Faculty of Agriculture and Forestry 4-year degree program

The onset of the 4-year degree program in the Faculty of Agriculture in 1977/78 academic session therefore provided an additional opportunity to give wildlife and fisheries management the prominence they deserve in the curriculum of the Department of Forest Resources Management. That was because of the popularity that the two programs enjoyed in the Faculty of Agriculture and Forestry. This development led to an opportunity for students to specialize in any of the following three areas in the final year:

Wildlife and range management

Inland fisheries management

Forestry

Thus, wildlife and fisheries management from this point constituted two-thirds of the undergraduate program in the Department of Forestry.

Founding the Department of Wildlife and Fisheries Management, University of Ibadan, Nigeria

Consequently, by the authority of university senate, the name of the Department of Forestry was changed to the Department of Forest Resources Management to reflect the additional mandates of wildlife and fisheries management.

This was essentially the position of wildlife and fisheries program in the university, until August 1, 1981 when, by the authority of university senate, the new Department of Wildlife and Fisheries Management was created.

From available records, the newly created department was the first full-fledged university functional Department of Wildlife and Fisheries Management on the continent of Africa. Therefore, it had a pioneering role to play in Nigeria, Africa, and the tropical world at large. Actually, the call for the creation of this department came from several sources. The UN/FAO at its various meetings called on the University of Ibadan in particular, to create a full-fledged university department of wildlife and fisheries management to serve the English-speaking countries of Africa. Calls also came at different times from the National Wildlife Conservation Committee, National Fisheries Development Committee, and the National Council for Agriculture. All the calls were in recognition of the pioneering efforts of the wildlife unit of the Department of Forest Resources Management on systematic studies of African wildlife and fisheries development.

Because of the basic facts leading to the creation of the Department of Wildlife and Fisheries, it became expedient to develop two separate curricula leading to full degree programs in wildlife management and fisheries management. It also became clear that the two programs must be "practical" in orientation without losing the required theoretical and all-embracing academic foundation to empower graduates as "job makers" rather than "job seekers" in consonance with the dictates of the economic situation in Nigeria and in Africa.

To achieve this objective, a committee comprising distinguished conservationists and fisheries and wildlife professionals from all over Nigeria was convened in November 1981 to prepare a plan of operation to realize this objective. By October 1982, the two undergraduate curricula leading to the degrees of BSc (Hons.) (wildlife management) and BSc (Hons.) (fisheries management) were approved by university senate.

Two sets of curricula in wildlife and fisheries emerged from the committee that worked closely with the newly recruited pioneering staff of the new department. The wildlife and fisheries curricula were in the mid-1982 taken physically by the author who was the pioneering head of the new department to University of Michigan at East-Lausing and Auburn University, Alabama, in the United States of America for vetting. This was to become the first Africa's indigenous full-fledged degree program in wildlife and fisheries management.

Both curricula covered the relevant aspects of wildlife and fisheries resources and operated within the 4-year degree program of the Faculty of Agriculture and Forestry. One year is fully devoted to practical work as in the case in the other departments within the Faculty of Agriculture and Forestry.

Finally, the university senate approved the comprehensive undergraduate and postgraduate curricula, that is, BSc and MSc and PhD degrees in wildlife and fisheries management, respectively, in October 1983.

Thus, within 2 years of its existence, the Department of Wildlife and Fisheries Management emerged with acclaimed undergraduates and postgraduate curricula, with full recognition for wildlife and fisheries as modern agricultural professions in Africa.

Students being trained in the department came from different parts of Africa, such as Ghana, Cameroun, Ethiopia, Zaire, and the Republic of Benin. It is interesting to note that, although the department of wildlife and fisheries management was initially mandated to serve the anglophone countries in Africa, it equally received students from French-speaking countries such as Zaire (now Democratic Republic of the Congo). In terms of staff strength, the department that started with only two academic staff, in 1969, had 22 full-time academic staff and 23 technical and administrative staff at the author's exit from university service in 2009. Student enrollment stood at 121 with 14 postgraduate MPhil/PhD students.

In summary, our teaching curriculum and research programs were designed to meet not only the needs of Nigeria but also those of the entire West and Central African countries that occupy a similar ecological zone. In pursuance of these programs, a number of linkages were established with reputable organizations and institutions within and outside Africa:

a. College of African Wildlife Management, Garoua, Cameroun
b. Department of Forestry and Wood Science, the Department of Animal Biology, and the School of Marine biology at the University College of North Wales Bangor, the United Kingdom
c. Department of Forestry and Natural Resources, University of Edinburgh, Scotland
d. Department of Biological Sciences, Cambridge, the United Kingdom
e. Institute of Aquaculture, University of Stirling, Scotland

The linkage programs have proved to be very beneficial, particularly in the area of staff development.

Curricullum development and capacity building in forestry, wildlife and fisheries management

There is no way to describe our previous academic programs without giving some of the personal experiences that are considered to be relevant to the objectives of this book. This is because the author was one of the original core staff of the Department of Forestry, the last of the so-called "counterpart" indigenous staff that was recruited by UN/FAO under that scheme that was established as a regional center for capacity building for protected area management and forestry for Africa.

During his over four decades of service to the University of Ibadan and to Africa, the author was part of the development of two out of the three academic programs that emerged from the original forestry department and its offshoots. It is believed that the University of Ibadan is now at an opportune time and place to review the past experiences and the impact that has been made on sustainable protected area management and rural development. It is also important that the university as a pioneering institution review the shortfalls of these programs and provide guidelines that will meet the challenges of the 21st century. It is on this basis that the following accounts of the previous and present programs are presented in this chapter with a proposal for the millennium goals.

The University of Ibadan is the oldest University in Nigeria. The Department of Forestry (now the Department of Forest Resource Management), established in 1963, was the first of its kind in Africa. The Department of Forestry later gave birth to the Department of Wildlife and Fisheries Management in 1981. The author was the pioneer head of the new department, 1981–88. As far as it is known, it was also the first university department to take full responsibility for capacity building for Africa in these fields, by offering full degree programs in wildlife and fisheries management. Furthermore, the three degree programs that emerged from the original Department of Forestry have been replicated in many English-, French-, and Portuguese-speaking universities in Africa. The two departments, that is, Forest Resources Management and Wildlife and Fisheries Management, have since their inceptions produced high-level manpower in natural resource conservation for numerous countries that are in responsible positions in conservation of protected areas in many countries in sub-Saharan Africa.

(a) First generation curriculum 1963–76

Contents and focus:

The first-generation curriculum was produced in 1963 by the internationally recruited expatriate FAO staff that established the Department of Forestry. The main features of the academic programs were as follows: forest ecology; forest management; silviculture; wood science; and forestry policy, law, and administration. The principle of agroforestry was at experimental, primordial stage at that time.

Shortfalls in the first generation curriculum:

The academic program was devoted almost entirely to cellulose forestry, with little attention to nonwood forest products. The teaching staff were diverse within the forestry profession but not multidisciplinary within the context of renewable natural resources, biodiversity, and the environment. The curriculum was therefore not based on the principle of integrated approach to conservation of natural resources. Furthermore, it was a rather exotic academic program, which was not based on African ecosystems mainly because Africa's flora and fauna were largely unstudied up to this point. Only half of a three-unit (30 h) course or 15 h only per year were devoted to wildlife conservation. At that time, fisheries was not featured at all in the syllabus, although it was recognized as a vital component of natural resources. The knowledge of aquaculture and the general biology of African fish species was so limited that all that was known about African fish was "fish" as a biological specimen. There was little information on the reproductive and regenerative capacities and parameters of population dynamics that are so basic to productivity of fish and fisheries and wildlife. There was no knowledge of the environmental factors that influence fish production in aquaculture or fish and wildlife in their natural habitats.

(b) Second generation curriculum 1976–81

Contents and focus:

The second-generation curriculum emerged in 1976/77 academic year, when the course system was introduced to the University of Ibadan and the programs of the Faculty of Agriculture and Forestry were transformed from 3 to 4 years. The course system allowed for flexibility and specialization as the curriculum became broad-based and more relevant to African situation than its predecessor. Another characteristic of this phase of development was the introduction of a "practical year program" that enabled undergraduates to spend a full year in the field during which they were exposed to practical realities of their trade. Furthermore the practical year prepared undergraduates to be practical "farmers" and "foresters" in total divergence from the colonial and postcolonial paths of producing "white-collar agriculture officers" and "forest officers." The latter category was less relevant to agriculture and forestry problems of Nigeria and Africa.

It is important to note that the contribution of previous research efforts, which started in the mid-1960s, permeated the second and third phases of curriculum development and capacity building. It was realized that research is a cornerstone upon which future management will be based. Results of research efforts were therefore constantly plowed back to teaching and curriculum development. Early research efforts were based toward inventory, taxonomy, and studies of the ecosystem characteristics in order to provide basic techniques and baseline information for the potential and management of renewable natural resources.

The 1976–81 curricula were consequently based on the following principles:

No renewable natural resources are inexhaustible, particularly when subjected to irrational exploitation and mismanagement. There was strong emphasis on "preservation" rather than "conservation" of what is left of Africa's natural resources.

For the first time, the curricula focused on an integrated, multidisciplinary approach to natural resources management; it emphasized the integrated approach to sustainable management of tropical ecosystems.

Utilization of indigenous natural resources as cultural, medicinal, and economic assets also became prominent. That was why emphasis was placed on "domestication" of our indigenous plants and animals that provide alternatives to conventional sources of food and medicine.

For the first time, the teaching of forestry, wildlife, and fisheries ecology featured prominently in the curricula based on the knowledge of the indigenous resources and ecosystems. These were direct consequences of pioneering research efforts that started in phase 1 of the curriculum development and that has continued to date. The academic program thus led to a degree structure with three options in forestry, wildlife, and fisheries. This phenomenon led to a change of name of the Department of Forestry to Forest Resources Management in 1976.

Short-falls in the second generation curriculum:

While the curriculum recognized the interdependence and interrelationship between "sectors" of natural resource, it did not embrace the holistic approach to the teaching and conservation of biodiversity. The modern approach is to study the totality of what constitutes the ecosystem and their interrelationships and not just the genes and species that are judged to be of economic importance today.

The curriculum was almost entirely based on the principle of preservation as a management strategy. The management tool was however meant for the welfare of the natural resources only and not for human welfare and rural development. The curriculum did not address the basic problem of poverty among rural dwellers who reside in closest proximity to natural resources. Furthermore, the curriculum did not prepare graduates to create structures and systems to which they will fit in and employ themselves and others on graduation from the university. In other words, it did not prepare university graduates to be self-reliant and self-employed.

There are demand-driven forces for curriculum development and capacity building that will make graduates more relevant, broad-based, and well consolidated upon the philosophy of self-employment, self-reliance, and job creation in natural resource conservation. Indeed, future efforts in capacity building should be so pragmatic and viable that protected area managers and instructors can confidently inform new undergraduates from their day 1 in the university that they are not meant for government jobs, because there may be none. Nevertheless, it can be said at the same time that they are assured that on graduation, they will be job creators and not job seekers, that is, they will be able to employ themselves and become self-reliant.

Lastly, the curricula did not embrace the principle of local community participation and utilization of natural resources for rural development. The academic programs should have highlighted the principle of integrated resource utilization as a means of expanding the economic base of local people who live in closest proximity to conservation and protected areas. The future curriculum should also highlight resource utilization as a tool for rural development. Future capacity building should therefore cater for rural communities as stakeholders and beneficiaries of professional programs and also address the problem of poverty in rural areas.

(c) Third generation curricula (1981–present)

Contents and focus:

The main focus of the forestry, wildlife, and fisheries curricula was the integrated approach to protected area management and conservation of natural resources. The broad-based curricula, which were also within the course system, at the University of Ibadan enabled undergraduates to select courses right across the three main spectra of natural resources—forestry, wildlife, and fisheries—and from the entire university that will enhance a full understanding of environmental protection and, at the same time, end up with specialized degrees in each of these sectors. Academic programs were impregnated with more indigenous knowledge of plant and animal species and the ecosystems that they inhabit. There was better understanding of surveys and inventory techniques and preparation of "management plans" and "master plans" for national parks and other protected areas. Natural resource economics, utilization of a wide range of natural resource, alternative medicine, and economic development; aquaculture, wildlife farming, and domestication of economic trees; population dynamics; and rational exploitation of natural resources were also the main features of the third-generation curricula.

Shortfalls in the third general curricula:

One significant omission in the third-generation curricula in forestry, wildlife, and fisheries was the lack of positive impact of previous efforts in capacity building on the national, subregional, and regional status of protected areas and natural resources in Africa. The declining and devastating status of African forests, the wildlife and fisheries, and the environment during these five decades of capacity building has been uninspiring, to say the least. This raises some fundamental questions and issues on what constitutes the vital ingredients for effective and sustainable utilization of our renewable natural resources such that it may well go beyond curriculum development and capacity building.

It must be emphasized that despite about five decades of capacity building in the management of protected areas, Africa's biodiversity is still full of unknowns, and our knowledge in these areas is still limited.

Impact of previous efforts and future academic and professional program

For capacity building to make a desired impact on the fast degrading environment in Africa, the following statements by the United Nations Environmental Programme (UNEP, 1992), may provide the platform for future metamorphosis of our programs:

> *There are clear indications that our knowledge of diversity on variability of plants and animals, microorganisms and the ecosystems in which they occur has become unbearably inadequate. This poses a collective threat to biodiversity.*
>
> ***Elizabeth DowdeswellExecutive Director, UNEP 1992***

> *The rural poor, which accounts for about 70% of Nigerian total population, depends almost entirely on biodiversity-related organisms for their survival and activities.*
>
> ***Nigerian Biodiversity Strategy and Action Plan (1998)***

These two statements suggest a dire need for the following changes and amendments in capacity building:

a. There must be a shift of emphasis from the narrow concept of natural resource conservation toward a holistic, total approach, to the conservation of biodiversity. Biodiversity spectrum includes all plants, all animals, all microorganisms, livestock, and their physical environment; air, water, and land; and their interrelationships.

b. Future programs should also focus on rational utilization of biodiversity for the welfare of man, for the rural poor, and for economic development.

c. There is need for a continuous and well-oriented problem-solving relevant multidisciplinary research programs into conservation and utilization of biodiversity and protection and improvement of the environment. It must be recognized, from historical perspective, that multidisciplinary research is the cornerstone upon which future resource management and environmental protection will be based.

Items (a–c) in the preceding text will provide a basis for diversifying the emergence of curricula that will include numerous professional programs in biodiversity-related themes. These concepts will define the role of African universities and other tertiary institutions and their relevance to the society, if capacity building is to have a significant impact on renewable natural resources. Other factors such as employment opportunities and self-reliance are discussed in the succeeding text:

The earlier analysis can provide the basis for future developments of programs that will embrace numerous professional programs in biodiversity-related themes. These programs can be at certificate and diploma levels. Emphasis will transcend techniques of conservation, to include techniques of utilization of biodiversity, biodiversity and business development, commerce, economics, administration, and conservation law. The program will place emphasis on scientific rational utilization of biodiversity for human welfare.

The emerging diploma and certificate courses will be targeted primarily to young school leavers, young graduates, and retired public servants, who wish to be self-employed and who wish to make a living from utilization of biodiversity. Since large expanse of land exists in rural areas, in closest proximity to biodiversity, land can be acquired at relatively low cost for polyculture of agriculture and natural resources for self-reliance.

CHAPTER 19

Utilization of biodiversity for entrepreneurship and job creation

Introduction

One of the indicators of the level of poverty in developing economics in Africa is unemployment. Studies indicate that unemployment has continued to grow in many parts of sub-Saharan Africa, particularly among the youth even in the face of several successive governments' multivarious efforts and expenditures aimed at confronting this challenge. In practical terms, it is important that various governments are committed to creating jobs, especially for the poor, comprising skilled and unskilled, school leavers, retirees, and rural dwellers. There is also an urgent need to address the barriers of time frames and access to funding and expected impacts of job creation schemes in Africa.

In the next decade or so, African countries will have to face unprecedented high level of unemployed young graduates of tertiary institutions and retirees from the public and private sectors. This is a significant category of the populace who on graduation and retirement have nothing to do, very little to eat, and nowhere to go. This is in combination with massive unskilled, unengaged population who live below poverty line in Africa. If the situation is allowed to linger on, the hardship, frustration, and other antisocial accompaniments of unemployment such as extreme poverty, hunger, and hopelessness will likely pose the biggest challenges to security and governance in African nations. That is why job creation in all sectors of the economy including natural resources and biodiversity conservation should be a priority of all tiers of governments and the private sector on the African continent.

A model for job creation

The agriculture and natural resources module displayed in this chapter can be adopted as a Mega Farm agroforestry enterprise for food production

Wildlife Conservation in Africa
https://doi.org/10.1016/B978-0-12-816962-9.00019-3

through innovation and new technologies for entrepreneurship, job creation, and poverty alleviation. It can also be adopted as part of skill acquisition centers in major ecological zones of Africa for promotion of capacity building.

The highlights of the scheme are applications of innovation and new technologies to boost food production and revenue generation. Furthermore the scheme will necessarily invite government intervention in terms of microcredits and revolving loans for establishment of small-scale business enterprises. The ultimate deliverable is job creation, food security, poverty alleviation, and rural development.

The following module can be used for job creation and entrepreneurship in utilization of biodiversity.

The solution to unemployment and poverty of African teeming youths will require collaborative and concerted innovative efforts by governments, nongovernmental organizations, and the private sector. Fig. 42 shows that agroforestry may well provide a key to successful entrepreneurship in the utilization of biodiversity. The nuclei of polyculture units such as grass cutter, giant rat, guinea fowl farming, fish farming, snailry, beekeeping, mushroom and vegetable farming, and an array of other possibilities can be sources of profitable business enterprises.

Possibilities for future development

It is suggested that in order to empower the unemployed youths in agriculture and biodiversity, capacity building or skill acquisition must take

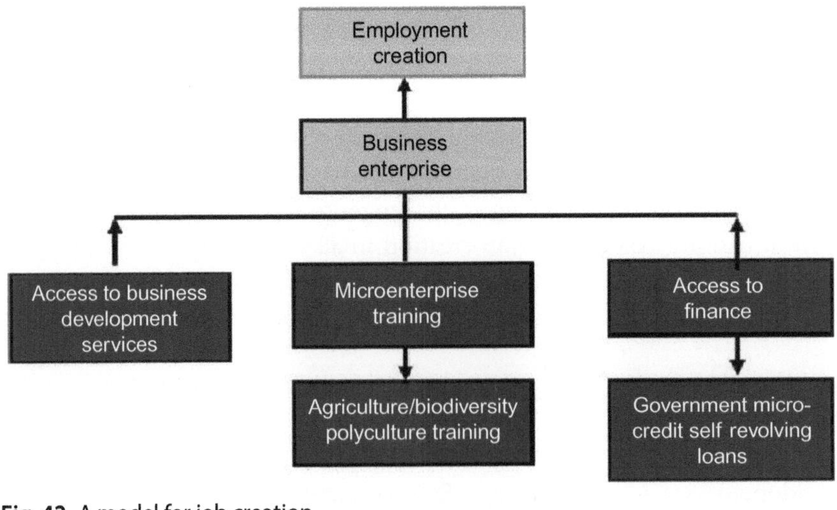

Fig. 42 A model for job creation.

precedence as the main element of entrepreneurship. Therefore the following training scheme can be integrated into the future curricula of forestry, wildlife, fisheries, and biodiversity in general. The outlines of the training program are presented in the succeeding text:

1. Food production with application of innovation and new technologies
2. Entrepreneurship skills and business management techniques for job creation and self-employment
3. Cooperative development and business management
4. Development of market structures and distribution of farm products
5. Packaging, preservation, transportation, and export of forest products
6. Promotion of local community participation in production method and transfer of production systems to rural areas
7. Mobilization of communities, project identification and prioritization, and development plans for acquisition of microcredits
8. Provision of technical skills for community participation in local government plans
9. Development of projects that are environmentally sustainable and that will contribute to the development of rural communities

Professional development of entrepreneurship in forestry, wildlife and fisheries

Training programs in the development of business enterprises from the chart of the nuclei of entrepreneurship presented on Table 48 are presented in the succeeding text:

Business creation and administration: A course in business creation will be designed to provide graduates of wildlife and fisheries management with the skills for creating and developing their own businesses. The program will foster entrepreneurial skills for small and big enterprises. Business administration will prepare wildlife and fisheries students for a career in business and will enhance their general learning skills and personnel development. It will enable students upon graduation to plan a start-up business without having to hire professional business administrators.

Personnel management: Personnel management is a vital business function that focuses on the performance of people within an organization to meet particular business objectives. This process involves human resource planning, recruitment, selection, orientation, training, performance appraisal, communication, compensation, and safety. It is expedient to educate the youths on these business elements, since the objective of modern program

Table 48 Establishment of a polyculture farm in agroforestry for training in entrepreneurship and job creation.

Establishment of a polyculture farm in agroforestry for training in entrepreneurship and job creation									
Main elements/commodities for entrepreneurship: biodiversity									
Forestry									
Agroforestry, agriculture wildlife farming, aquaculture, and forestry									
Herbal garden	Beekeeping	Mushroom farming	Snailry	Grass cutter farming	African giant rat farming	Guinea fowl farming	Fisheries and aquaculture	Common vegetables	Maize and cassava
Nuclei of enterprises for business promotion									
Herbal medicine	Wax honey	Mushroom	Meat	Bushmeat	Bushmeat	Meat	Fish	Vegetables	Stable foods

of wildlife and fisheries and biodiversity is to produce skilled graduates that can productively contribute to the development of the society.

Microenterprise training: The program of microenterprise training provides a small business training that envelopes such issues as how to write a business plan and obtain financing, accounting and tax issues, business insurance, legalities, marketing, and other basic business start-up issues. This program will also prepare the students for the task of business management.

Accounting: Students will need to understand the basic principles of accounting and financial management in order to improve the effectiveness of entrepreneurship skills. This sector will include framework of essential financial terminology, concepts, and applications that will enable entrepreneurs to analyze and interpret commonly used financial information in making business decisions. The program will enable students to grapple managerial accounting, understand the performance matrix of organizations, read and use company reports, etc.

Advertising: It is expedient for students to acquire the knowledge of advertising that will guide them toward successful business ventures. Advertising is the promotion of a product or service and is extremely persuasive in contemporary society. To maximize sales, companies will pay a premium for wide exposure through the mass media. Advertising space is common, but not restricted to these realms: social media, billboards, public transportation, movies (product placement), schools, and clothing; even bathroom stalls carry advertisements as the industry is constantly finding new ways to advertise.

Information technology and computer skills: The world is in the information age, a time when information is shared in a speed of light. Students will be exposed to information technology to learn the opportunities in this area of development. Mastering computer skills begins with acquiring a high typing speed followed by learning major office applications such as Microsoft Office Word, Excel, PowerPoint, Outlook, and Access. Utilization of these software programs effectively will open opportunities for most office jobs. Every student that wants to be exceptional in learning business enterprises in wildlife and fisheries production must, as of necessity, pay attention to these skills.

Logistic management: Logistics is that part of the supply chain that plans, implements, and controls the efficient, effective forward and reverse flow and storage of goods, services, and related information between the point of origin and the point of consumption in order to meet customers' and legal requirements.

Government and nongovernment interventions in job creation

Interventions of government in entrepreneurship and job creation will include the following:

It is envisaged that government intervention will begin with land acquisition, provision of all physical infrastructures, and cost of personnel for the training schemes. It will be a self-financing polyculture model.

Development of microcredit schemes and revolving loans to small-scale enterprises.

Development of cooperative/entrepreneurship societies.

Development of effective mechanisms for repayment of revolving loans.

CHAPTER 20

Model academic programs in wildlife and fisheries management*

Academic programs of the Departments of Wildlife and Fisheries Management, University of Ibadan

Undergraduate study

The current academic programs of the Department of Wildlife and Fisheries Management presented here represent the third generation of curricula that emerged from the original Forestry Department that was created in 1963 (Figs. 43 and 44). The first-year courses introduce students to basic principles of environmental biology. Most of these courses also apply to students of other Departments of the Faculty of Agriculture and Forestry, of University of Ibadan. The first-year courses also cover basic principles of forestry, fisheries, and wildlife. The curriculum also covers the basic biology of the three principal resources and other basic requirements for students of agriculture. The second-year courses provide the theoretical introduction to all the principal branches of forest resources management. However, genetics, crop physiology, and forest entomology and pathology are taught by the Department of Agricultural Biology, while the remaining courses are provided by the Department of Wildlife and Fisheries Management. These comprise basic techniques in resource inventory, data processing, and analysis; studies of ecology, tree crop production, and wood science; and principles of economics and management of forestry, fisheries, and wildlife. An introduction to land use planning is given, and the problems of access to the resource and harvesting methods are described. The full list of course titles is given in the succeeding text.

*Note that this is the first curriculum that was introduced to teach wildlife and fisheries management in Africa, and this can change over time. However, this curriculum template can be used as a foundation for curriculum development in wildlife and fisheries management in other Universities across sub-Saharan Africa and can be improved upon with time.

Wildlife Conservation in Africa
https://doi.org/10.1016/B978-0-12-816962-9.00020-X

Fig. 43 The Department of Forestry, University of Ibadan, Nigeria, as the author met it when he took up his first appointment as a temporary assistant training fellow in 1969. It was the building where 1½ units of wildlife courses were domiciled in 1969.

The third year is devoted to practical studies. A large proportion of instruction time is spent in the field. Practicals are however backed by theoretical courses. By the end of the third year, students would have acquired a sound working knowledge of the subject of forest resources management in general. Up to this stage, the theme of the course is to stress the essential unity of the study of renewable natural resources, that is, wildlife, fisheries, and forestry, and to provide a background that may lead equally to any of the specializations. In the final year, 50% of the courses are taken by all students. The other half of the final year courses enable specialization in any of the five subjects. They are forest biology, silviculture, forest economics and management, wood science, fisheries management, and wildlife and range management.

Postgraduate study

The master of science (MSc) program extends over a full year, from October to September, and consists of three terms of lectures and practicals with examinations followed by a 3-month period for a study project and dissertation. Admission is normally open to university graduates in forestry, agriculture, the biological sciences, or any other relevant disciplines. This program is an advanced training course for specialist workers. The master

Fig. 44 The Department of Wildlife and Fisheries Management when the author retired at a mandatory retiring age of 65 years in 2009 where the current two separate degree programs of wildlife and fisheries are domiciled.

of philosophy (MPhil) requires a minimum of six terms of full-time study or nine terms for part-time candidates. Further course work and examinations are required, but the program is mainly devoted to work on a research project leading to an original dissertation. Direct admission is limited to holders of a second class (upper division) honor degree or higher qualification, but a candidate who has satisfied the requirements for MSc may apply for permission to transfer to MPhil.

The doctor of philosophy (PhD) requires a minimum of nine terms of full-time study or 12 terms of part-time study. Course work and examinations are required, but the principal examination is based on the thesis that must show evidence of the candidate's ability to draw significant conclusions from his own research. Admission to the course is normally by transfer from the MPhil.

Each of the higher degrees thus requires an initial period of study, normally lasting 1 year. At the end of this year, the two research degrees MPhil and PhD may be completed after 1 or 2 further years of full-time work, respectively, but they may be undertaken part-time with a corresponding increase of the minimum period. It is therefore possible for an officer in the administration or in a research institute to return to his base after a year's study leave and continue to work for the degree.

MSc courses will be available in six areas of specialization:

Forest biology and silviculture

Forest biometrics and remoter sensing

Forest economics and management

Fisheries management

Wildlife and range management

Wood science

A number of "core courses" including such subjects as statistics in resources management, organizing research, and preparation of a seminar are compulsory for all candidates. Each of the six specialist areas, however, offers more courses than the required minimum standard so that candidates may make a further selection from within their chosen area. It is also possible for candidates to attend and receive a credit for courses at postgraduate level in other departments. Courses within the Departments of Forest Resources Management and the Department of Wildlife and Fisheries are presented in the succeeding text.

Course acronyms

The following are course "codes" or acronyms used for designation of courses in the Department of Forest Resources Management and the Department of Wildlife and Fisheries Management.

GES—General studies

AGE—Agricultural economics

PHY—Physics

CHE—Chemistry

ZOO—Zoology

BOT—Botany

CPE—Crop protection and environmental management

AGY—Agronomy

HED—Home economics and development

WFM—Wildlife and fisheries management

ANS—Animal science

FRM—Forest resources management

ABN—Animal biochemistry and nutrition

VAN—Veterinary and agricultural nutrition

SOS—Soil science

AED—Agricultural extension and rural development

FIS—Fisheries studies

WFPYTP—Wildlife, fisheries, practical training

WRF—Wildlife, range and fisheries

VN—Veterinary nutrition

Key to acronyms used for registration of courses

C="Compulsory" courses that are domiciled within the department. All courses designed as "C" are compulsory for all students, that is, all students must register for them and score at least 40%.

R="Required" courses are those relevant to students' profession, but are not captured within the category of "compulsory courses." Students who register for them must score a minimum of 30%. Where there is a group of such courses, that is, 2 or 3 or more, the department will specify the minimum number of units that students must register for.

E="Elective" courses are those relevant to the profession of the students that are not captured in the category of "compulsory" and "required" courses. Elective courses enable students to make up the additional units for the award of his/her degree. Students are however not compelled to pass it. However, it becomes an additional bonus if they do well in elective courses.

Current courses in the Department of Wildlife and Fisheries Management, University of Ibadan

Wildlife management undergraduate courses.

100 Level			
Course code	**Course title**	**Units**	**C/R/E**
GES 101	Use of English	3	C
GES 106	Philosophy and logic	3	C
AGE 110	Mathematics in agriculture	3	C
PHY 101	Physics for agriculture, forestry, and veterinary medicine	4	R
CHE 157	Physical chemistry	4	R
CHE 127	Inorganic chemistry	3	R
ZOO 112	The mammalian body	3	R
CHE 177	Organic chemistry	3	C
CHE 195	Practical chemistry	2	C
ZOO 111	Principle of animal biology	3	C
ZOO 113	Diversity of animal life	3	R
BOT 141	Basic principles in botany	3	C
BOT 121	Introductory biology of phanerogams	3	E
BOT 111	Cryptogamic botany	3	E
Total no. of units (C=20; R=17; E=6)			

Continued

Wildlife management undergraduate courses—cont'd

200 Level			
Course code	**Course title**	**Units**	**C/R/E**
AGE 201	Introduction to computer science	2	2
GES 103	Government society and economy		3
AGE 210	Introduction to agric. and economics	2	2
CPE 210	Introduction to crop and forestry botany	3	C
AGY 210	Introduction to general agriculture (crop)	2	C
HED 210	Introduction to home economics	3	C
*WFM 211	Introduction to West African fish and wildlife	3	C
ANS 210	Introduction to gen. agric. (livestock)	3	C
FRM 210	Introduction to forest resources mgt.	3	C
ABN 200	Introduction to agricultural biochemistry	2	R
VAN 225	Anatomy and physiology	3	R
AGY 212	Introduction to agro-climatology	2	R
SOS 210	Elements of soil science	2	R
GPE 211	Introduction to plant protection	2	R
ABN 220	Introduction to food science and tech.	2	R
AED 210	Introduction to agric. extension	2	C
Total no. of units (C = 23; R = 15)			

300 Level			
Course code	**Course title**	**Units**	**C/R/E**
WFM 311	Wildlife genetic and animal breeding	2	R
WFM 312	Metabolic interrelationship in fish and wildlife	2	C
WFM 313	Introduction to tourism	2	C
WFM 314	Ecotourism in protected area	2	C
WRM 320	Wildlife population analysis	2	C
WRM 321	Wildlife ecology and management	2	C
WRM 322	Principles of wildlife and range mgt.	2	C
WRM 323	Principles of conservation in tropical Africa	2	R
WRM 324	Principles of park planning mgt. and admin.	2	R
WRM 325	Wildlife pest and diseases	2	R
WRM 326	Wildlife business mgt. and marketing	2	R
WRM 327	Wildlife and fisheries environment mgt.	2	R
WRM 328	Herpetology	2	R
WRM 329	Geographic survey and information practice	2	R
FRM 320	Forest biometrics I	2	C
FRM 319	Forest operation I	2	R
FRM 318	Introduction to land use planning	2	R
FRM 315	Forest economics	2	C

Wildlife management undergraduate courses—cont'd

300 Level

Course code	Course title	Units	C/R/E
FRM 312	Natural ecosystems	3	R
FRM 311	Resource inventory and menstruation	3	R
AED 310	Introduction to agric. extension and rural sociology	2	R

Total no. of units (C = 16; R = 25)

400 Level

Course code	Course title	Units	C/R/E
WRM 420	Wildlife ecology survey	2	C
WRM 421	Wildlife stock assessment	2	C
WRM 422	Laboratory analytic technology in wildlife 2	2	C
WRM 423	National parks and game reserve mgt.	2	C
WRM 424	Training in firearms and ballistics	3	C
WRM 425	Park planning and tourism	3	C
WRM 426	Road and bridge construction in wildlife reserve	3	C
WRM 427	Zoo management and museum techniques	3	C
WRM 428	Animal capture techniques	3	C
FIS 427	Pond construction and management	3	C
FRM 417	Forest biometric II	2	C
WFPYTP 420	Industrial attachment	6	C

Total no. of units (C = 34; R = 0)

500 Level

Course code	Course title	Units	C/R/E
WRM 518	Seminar	2	C
WRM 519	Special project	4	C
WRM 521	Wildlife administration, legislation, and conservation	2	C
WRM 522	Wildlife utilization	2	R
WRM 523	The concept of park and zoo management 2	2	R
WRM 524	Game ranching and domestication	2	C
WRM 525	Management of game birds	2	2
WRM 526	Wildlife extension and education	2	2
WRM 527	Wildlife nutrition	2	C
FRM 514	Natural resources policy and admin.	2	R
FRM 511	Multiple land use	2	R

Total no. of units (C = 12; R = 12)

Fisheries management undergraduate courses.

100 Level			
Course code	**Course title**	**Units**	**C/R/E**
GES 101	Use of English	3	C
GES 106	Philosophy and logic	3	C
AGE 110	Mathematics in agriculture	3	C
PHY 101	Physics for agriculture, forestry, and veterinary medicine	4	R
CHE 157	Physical chemistry	4	R
CHE 127	Inorganic chemistry	3	R
ZOO 112	The mammalian body	3	R
CHE 177	Organic chemistry	3	C
CHE 195	Practical chemistry	2	C
ZOO 111	Principle of animal biology	3	C
BOT 141	Basic principal in botany	3	C
BOT 121	Introductory biology of phanerogams	3	E
Total no. of units (C = 20; R = 14)			

200 Level			
Course code	**Course title**	**Units**	**C/R/E**
AGE 201	Introduction to computer science	3	C
AED 210	Introduction to agric. extension	3	C
GES 103	Government society and economy	2	C
WFM 211	West African fish and wildlife	3	R
ANS 210	Introduction to genetics in agriculture (crop)	2	R
CPE 210	Introduction to crop and forestry botany	3	R
AGY 210	Introduction to general agric. (crop)	2	R
HED 210	Introduction to home economics	3	C
FRM 210	Introduction to forest resource mgt.	2	R
ABN 200	Introduction to agric. biochemistry	3	R
VAN 225	Veterinary anatomy and physiology	3	R
CHE 259	Physical chemistry and live science	2	R
CHE 279	Basic aromatic and natural product chemistry	3	R
AGY 212	Principles of climatology	3	R
SOS 210	Elements of soil science	2	R
AGE 212	Introduction to agric. economics	2	R
ABN 220	Introduction to food science and tech.	2	R
Total no. of units (C = 13; R = 29)			

Fisheries management undergraduate courses—cont'd

400 Level			
Course code	**Course title**	**Units**	**C/R/E**
FIS 420	Fishing gear use, design, production, and maintenance	2	C
FIS 421	Fisheries stock assessment	2	C
FIS 422	Fisheries recreational planning and interpretation	2	C
FIS 423	Fish production management technology accounting practices	3	C
FIS 424	Oceanographic techniques	2	C
FIS 426	Fisheries engineering	2	C
FIS 427	Pond construction and management	3	C
FIS 428	Techniques in fish processing and utilization	2	C
FIS 429	Aquatic ecological survey	2	C
FIS 430	Analytic techniques in fisheries	3	C
WRM 427	Zoo management and museum techniques	3	C
WFPYTP 429	Industrial attachment	6	C
Total no. of units (C = 35; R = 0)			

500 Level			
Course code	**Course title**	**Units**	**C/R/E**
WRF 518	Seminar	2	C
WRF 519	Special project	4	C
FIS 520	Ornamental fisheries and aquaria design	2	R
FIS 521	Fish production and management	2	C
FIS 522	Fish production dynamics	2	C
FIS 523	Advanced fish nutrition	2	R
FIS 524	Fisheries economics	2	R
FIS 525	Farm management and fisheries business management	2	C
FIS 526	Fish technology processing and storage	2	R
FIS 527	Water quality management and pollution control	2	R
FIS 528	Production of other marine products	2	R
FIS 529	Nigerian feeds and feeding stuffs	2	R
FIS 530	Fish farming engineering	2	R
FIS 531	Administration and program planning in ext.	2	R
FIS 532	Fisheries policy and legislation	2	R
Total no. of units (C = 12; R = 20)			

Continued

Forest resources management undergraduate courses.

100 Level

Course code	Course title	Units	C/R/E
GES 101	Use of English	3	C
GES 106	Philosophy and logic	3	R
CHE 127	Inorganic chemistry	3	R
CHE 157	Physical chemistry	4	R
CHE 177	Organic chemistry 1	3	C
CHE 195	Practical chemistry	2	C
BOT 111	Cryptogamic botany	3	R
BOT 121	Introductory biology of phanerogams	3	R
BOT 141	Basic principle in botany	3	C
ZOO 111	Principle of animal biology	3	C
ZOO 113	Diversity of animal life	3	R
PHY 101	Physics for agric. forestry and veterinary medicine	4	R
AGE 110	Mathematics for agriculture	3	C
ZOO 112	The mammalian body	3	R
Total no. of unit (C = 20; R = 20; E = 3 = 40)			

200 Level

Course code	Course title	Units	C/R/E
FRM 210	Introduction to forest resources mgt.	3	C
GES 103	Government society and economy	2	C
AGY 210	Introduction to gen. agric. (crop)	2	C
AGY 212	Introduction to agro-climatology	2	R
SOS 210	Elements of soil science	2	R
CPE 210	Introduction to crop and forestry botany	3	C
CPE 211	Introduction to plant protection	2	R
CPE 213	Crop anatomy and physiology	2	R
AGE 210	Introduction to agricultural economics	3	C
ABN 200	Introduction to agricultural biochemistry	2	R
ABN 220	Introduction to food science and tech.	2	R
ANS 220	Introduction to gen. agric. (animals)	2	C
AGE 201	Introduction to computer science	2	R
WFM 211	Introduction to West African fish and wildlife	3	C
AED 210	Introduction to agric. extension	2	C
AED 210	Introduction to home economics	3	R
VN 225	Veterinary anatomy and physiology	3	R
Total no. of units (C = 21; R = 20) = 41			

Forest resources management undergraduate courses—cont'd

300 Level			
Course code	**Course title**	**Units**	**C/R/E**
FRM 311	Resources inventory and mensuration	2	C
FRM 312	Natural ecosystems	3	C
FRM 313	Wood: its formation and properties	2	C
FRM 314	Principles of silviculture	3	C
FRM 315	Forest economics 1	2	C
FRM 316	Forest management 1	2	R
FRM 318	Introduction to land use planning	2	R
FRM 319	Forest operation 1	2	R
FRM 320	Forest biometric 1	3	C
FRM 321	Introduction to urban forestry	2	2
FRM 322	Introduction to chem. processing of forest produce	2	R
FRM 323	Ethnoforestry 1	2	R
FRM 324	Forestry taxonomy	2	R
FRM 325	Introduction to environ. impact assessment	2	E
CPE 310	Introduction genetics	3	R
CPE 320	Principles of forest protection	2	R
WFM 311	Wildlife ecology and management	2	E
WFM 313	Prin. of conservation in tropical Africa	2	E
WFM 314	Principles of park planning management administration	2	E

Total no. of units (C – 16; R = 22, E – 8) – 38

400 Level			
Course code	**Course title**	**Units**	**C/R/E**
FRM 411	Elementary ground survey	2	C
FRM 412	Forest inventory and mgt. plan.	2	C
FRM 413	Aerial survey and photogrammetry	2	C
FRM 414	Practical silviculture	3	C
FRM 415	Harvesting and utilization of forest product	3	C
FRM 416	Ecological survey	2	C
FRM 417	Forest biometric II	3	C
FRM 418	Forest operation II	3	C
FRM 420	Communication skills in renewable natural resources	2	C
FRM 421	Entrepreneurial skills in renewable natural resources	3	C
FRM 422	Agro-forestry systems	2	C
FRM 400	Practical approaches to forestry practices (SIWES)	6	C

Total units (C = 33) = 33

Continued

Forest resources management undergraduate courses—cont'd

500 Level			
Course code	**Course title**	**Units**	**C/R/E**
FRM 511	Multiple land use	2	C
FRM 512	Forest protection and conservation	2	C
FRM 513	Forest economics II	2	C
FRM 514	Natural resources policy law and admin.	3	C
FRM 515	Forest management II	2	R
FRM 516	Forest mensuration	2	R
FRM 517	Forest soil	2	R
FRM 518	Biological control of timber quality	2	C
FRM 519	Forest genetics and three improvement	2	C
FRM 520	Special project	2	C
FRM 521	Tropical tree crop production practices	4	R
FRM 522	Production ecology	2	E
FRM 523	Forest industry and marketing	2	E
FRM 524	Processing of forest produce	2	E
FRM 525	Preservation of forest produce	2	E
FRM 526	Extension and community development in renewable natural resources	2	C
FRM 527	Ethnoforestry II	2	R
Total no. of units (C = 17; R = 10; E = 8) = 27			

Suggestions for enrichment of current academic programs in protected area management

The executive secretary of the United Nations Environmental Programme (UNEP) in 1992 declared that basic knowledge of biodiversity is scanty. This is one of the fundamental reasons for the loss and underdevelopment of biodiversity particularly in tropical Africa. Furthermore, in consonance with "Agenda 21" of United Nations Environmental Programme (UNEP, 1992) blueprint for the development of biodiversity in the 21st century, the new horizon for capacity building in biodiversity will therefore embrace the following themes:

1. Basic empirical knowledge of plants, animals, microorganisms, air, water, and land that together constitute biodiversity.
2. Other areas of omissions or weak emphasis in current programs that must be addressed in capacity building in biodiversity in consonance with the spirit of "UNEP Agenda 21 blueprint" are plant and animal taxonomy. The knowledge of identification and classification of plants and animals in different ecological and geographic zones of tropical Africa is fundamental

to the development of biodiversity. The anatomy, physiology, and basic management principles of species—threatened, unthreatened, and endemic—are basic to the survival of biodiversity.

3. The structure, characteristics, and productivity of major tropical ecological zones in Africa—the mangrove, the tropical rain forest, the African savanna, and the Sahel regions of Africa.
4. Management of biodiversity on fragile ecosystems and marginal environments such as arid and semiarid lands, erosion-proned ecosystems, rugged terrains, and regions that can be classified as productive for arable farming.
5. Management of special tropical ecosystems such as aquatic and semiaquatic environment conservation of wetlands and its significance to migratory species of animals—particularly, the palearctic, agro-tropical migratory bird species.
6. Soil characteristics and soil types in relation to plant and animal associations and plant and animal communalities.
7. Human activities and loss of biodiversity: exploitation, overexploitation of biodiversity, pollution of the environment, mismanagement of the environment management strategies that are themselves problematic to sustainable use of biodiversity, desert encroachment, agents of erosion, and loss of biodiversity. Burning regimes that result in loss of biodiversity, extensive pastoralism, and grazing regimes that cause environmental degradation.
8. Management strategies for restoration of biodiversity, principles of environmental protection, restoration of degraded polluted environments, shelter-belt afforestation for reclamation of arid and semidesert environments, afforestation for different kinds of land use, and afforestation as a tool for restoration of biodiversity.

Holistic conservation of biodiversity for human welfare, food, medicare, goods and services as a strategy for future curriculum development

The outlines for the holistic management of biodiversity are presented in the succeeding text:

a. Resources planning: survey techniques, data processing, plants and animal enumeration, animal census, and basic principles of formulation of management plans.
b. Protected area management—to maintain the status quo for perpetuity.
c. Principles of rational exploitation of biodiversity and sustainable use.
d. Strategies for reconciling the primary objectives of conservation of biodiversity with its utilization for human welfare. Utilization for human welfare such as food, medicare, goods, and service.

e. "System plan" as a tool for reconciling management objectives, economic principles for conservation of biodiversity, and land use and management plan for conflict resolution.
f. Basic national and international requirements for the development of policies for conservation of biodiversity. Policy reform for conservation of biodiversity and environmental protection.
g. Policy, law, and administration in conservation of biodiversity. Basic components of forest and wildlife legislation. Law enforcement in the conservation of biodiversity.
h. Bringing the benefits of conservation to rural communities. The principle, concept, and elements of local community participation in the conservation of biodiversity. The elements of CAMPFIRE, Communal Area Management Programme for Indigenous Resources in Zimbabwe, and LIRDP, Luangwa Integrated Rural Development Programme in Zambia.
i. International conventions and their relevance to global conservation of plants and animal species, environmental protection, migratory animals, etc.

References

Ajayi, S.S., 1974. The Biology and Domestication of the African Giant Rat (*Cricetomys gambianus Waterhouse)* (Ph.D. thesis). University of Ibadan.

Ajayi, S.S., 1975. Domestication of the African Giant Rat. Kwara State Publishing Co., Ilorin, Nigeria.

Ajayi, S.S., 1976. An Approach to the Domestication of African Rodents. F.A.O. Working Party on Wildlife Management and National Parks, 5th session 1976. FAO, Rome, Italy, 8 pp.

Ajayi, S.S., 1978. The utilization of tropical forest wildlife: state of knowledge and research priorities. In: Invited Special Paper Eight World Forestry Congress, Jakarta Indonesia, 16–28 October 1978, pp. 1–3. Ref. FFW/8-6.

Ajayi, S.S., 1979. Utilization of Forest Wildlife in West Africa. FO: Misc/79/26, FAO, Rome, Italy, 76 pp.

Ajayi, S.S., 1981. Meat production potential of some recently domesticated wildlife species. A report to the Nigeria Committee on Green Revolution.

Ajayi, S.S., 1990a. Training seminars on integrated wildlife resource use. Arrangements for SADCC training seminars on integrated wildlife resource use.

Ajayi, S.S., 1990b. Training seminars on integrated wildlife resource use. Record of SADCC mobile training seminars on wildlife managements involving people's participation.

Ajayi, S.S., 1990c. Training seminars on integrated wildlife resource use: rural community participation in integrated wildlife management and utilization in Botswana, Zambia, and Zimbabwe – A collection of seminar papers.

Ajayi, S.S., Tewe, O.O., 1978. Performance of the African giant rat (*Cricetomys gambianues Waterhouse*) on the commercial rations and varying dietary protein levels. Lab. Anim. J. 12, 109–112.

Ajayi, S.S., Tewe, O.O., 1979a. The Proximate and Mineral Composition of Smoked Meat of Some Forest Mammals in Southern Nigeria. p. 34.

Ajayi, S.S., Tewe, O.O., 1979b. Replacement and value of some energy substitute for the African giant rat (*Cricetomys gambianus waternouse*). J. Inst. Anim. Tech. 31 (1), 15–23.

Ajayi, S.S., Tewe, O.O., 1980. Food preference and carcass composition of the grasscutter (*Thrynomys swiderianus Termick*) in captivity. Afr. J. Ecol. 18, 133–140.

Ajayi, S.S., Halstead, L.B., Geerling, C., 1979. Wildlife Management in Savannah Woodland. Taylor & Francis, Ltd.

Ajayi, S.S., Oguntala, B.A., Adedipe, N.O., Osuntogun, N.O., Okojie, J.E., Fubara, M.O., 1998. Nigeria Biodiversity Strategy and Action Plan. World Bank/Federal Government of Nigeria Environmental Protection Agency, Abuja, Nigeria.

Akum, Z.E., 1979. Economic Importance of Wildlife in Cameroon. (Unpublished Ms).

Ashford, R.W., 1970a. A possible reservoir for Leishmania tropica in Ethiopia. Trans. R. Soc. Trop. Med. Hyg. 64 (6), 936–937.

Ashford, R.W., 1970b. Observations on the biology of Hemimems talpoides (*Insecta: Dermaptera*). J. Zool. 162 (3), 413–418.

Asibey, E.O.A., 1970. Shai, Hills Bush Meat Production Project. Unpublished Ms, Department of Game and Wildlife, Accra, Ghana. pp. 21, 26, 28.

Asibey, E.O.A., 1974a. The grass cutter (*Thryonomys swinderianus Temminck*) T, Ghana. Symp. Zool. Soc. Lond. 34, 161–170.

Asibey, E.O.A., 1974b. Wildlife as a source of protein in Africa South of the Sahara. Biol. Conserv. 6 (1), 33–39.

Asibey, E.O.A., 1976. Primate conservation in Ghana. In: Chivers, D.J., Petter, W.L. (Eds.), Recent Advances in Primatology, vol. 2: Conservation. Academic Press, London, New York, San Francisco, pp. 55–74.

Asibey, E.O.A., 1978. Wildlife production as a means of meat supply in West Africa, with particular reference to Ghana. In: Paper presented at 8th World Forestry Congress, Jakarta, October 16–28, 1978. FF-F/8-5 (21 pp.).

Asibey, E.O.A., Eyeson, K.K., 1975. Additional information on the importance of wild animals as a food source in Africa South of the Sahara. J. Ghana Wildlife Soc. 1 (2), 13–18.

Baldus, B.D., Cauldwell, A.E., 2004. Tourist Hunting and Its Role in Development of Wildlife Management Areas in Tanzania. Games Management Department for Sub-Saharan Tanzania.

Barnes, D.K., Galgani, F., Thompson, R.C., Barlaz, M., 2009. Accumulation and fragmentation of plastic debris in global environment. Philos. Trans. Royal Soc. B264, 1985–1998.

Bates, G.L., 1905. Notes on the mammals of Southern Cameroon and Benito. Proc. Zool. Soc. Lond., (1), 68–85.

Bigalke, R.C., 1964. Can Africa produce new domestic animals? New Scient. 374, 141–146.

Booth, A., 1959. On the mammalian fauna of the Accra Plains. J. West Afr. Sci. Ass. 5, 26–36.

Booth, V., 2002/2009. Contribution of Wildlife to National Economies – CIC Technical Series No. 8: Analysis of Wildlife Markets (Sport hunting and Tourism), WWF (SARPO – Harare).

CEC, OECD, VN and WTO, 2001. Tourism Satellite Account: Recommended Methodological Framework. Commission for the European Communities, Brussels; Eurostat, Luxembourg; Organization for Economic Co-operation and Development, Paris; United Nations and Madrid World Tourism Organization, New York, p. 137.

Charter, J.R., 1970. The economic value of wildlife in Nigeria. In: Proc. Forestry Association of Nigeria, Annual Conference, Ibadan. 13 p.

Child, G.S., 1979. Report on the establishment of national parks and reserves in Libya. The Libyan Arab Jamahirya. 11 pp., March 27.

Child, B.A., 2002. Background Assessment of the Wildlife Sector, Zambia. Norwegian support to the environmental management sector with main focus on wildlife management, Lasaka, Zambia. Draft Report, p. 52.

Coe, M.J., Cumming, D.H.M., Philipson, J., 1976. Biomass and production of large herbivors in relation to rainfall and primary production. Ecologia 122, 341–354.

Damm, G., 2005. Hunting in South Africa: facts, risks and opportunities. African Indaba 2, 1–14.

Darling, F.F., 1960. Wildlife in an African Territory, vol. VII. Oxford University Press, London. p. 160.

Day, G.M., Linn, I., 1972. Notes on the food of feral mink (mustella vision) in England and Wales. J. Zool. Lond. 162, 419–425.

De Leeuw, P.N., 1977. Animal nutrition and livestock development in the Nigeria savanna region. In: Conference on the Aftermath of the Drought in Nigeria. Ahmadu Bello University, Zaria.

Egoscue, H.J., 1972. Breeding the long-tailed pouched rat, Beamys hindei, in captivity. J. Mammal. 53 (2), 296–302. https://doi.org/10.2307/1379165.

Elley, T.J., 1970. Stomach contents, weights and volumes of cape hare. E. Afr. Wildl. J. 8, 202.

Erb, J., 2001. Population dynamics of large and small mammals. Oikos 92 (1), 3–12.

Essang, S.M., Olagide, S.O., 1974. Intermediate Economic Analysis, first ed. Aromolaran Publishing Limited, Ibradan, Nigeria. pp. 125.

FAO, 1967. Trade in wild animals and their products. FAO Secretariat Note, In: Ad-Hoc Working Party on Wildlife Management, Fort Lamy, Chad, 6–11 February 1967.

Federal Department of Forestry—Nigeria, 1977. Progress report on development of Nigeria's wildlife resource. In: 9th Meeting, National Council for Agriculture and Rural Development, Ilorin, 8–9 February 1978. Federal Department of Forestry. Unpublished Ms. 9 pp.

Field, C.R., 1972. The food habits of wild ungulates in Ugandan by analysis of stomach contents. E. Afri. Wildl. J. 10, 1–16.

Fisher, J.L., 1968. National Resources Wise Use of the World's Inheritance. Environmental and Policy, Indiana University Press, USA.

Gabrielson, I.N., 1970. Wildlife Conservation. Macmillan, London.

Gambino, J.J., Lindberg, R.G., Hayden, P., 1965. A search for mechanisms of radiation resistance in pocket mice. Radiat. Res. 26 (2), 305–317.

Hatt, P.K. and Reiss, A. J. (eds) (1940). Am. J. Sociol. XLVI: 191 – 226.

Hill, W.C.O., Porter, A., Bloom, R.T., Seago, J., Southwick, M.D., 1955. Field and laboratory studies on the naked rat heterocephalus glaber. Proc. Zool. Soc. Lond. 128, 455–513.

I.U.C.N., U.N.E., W.W.F, 1978. A World Conservation Strategy. IUCN, Switzerland.

IUCN, 1998. Directory of Afro Tropical Protected Areas. IUCN, Gland. 1034 pp.

Kenward, R., Sharp, R., 2008. Use nationally of wildlife resources across Europe. In: Manos, P., Papathanasious, J. (Eds.), GEM-CON-BIO: Governance and Ecosystems Management for the Conservation of Biodiversity, pp. 117–132. Thesaloniki. Funded by the 6th Framework Programme and co-financed by the European Commission with contract No: 028827.

Kramlich, W.E., Pearson, A.M., Tauber, A., 1973. Process Meat. The AVI Publishing Co. Inc., Westport, Connecticut.

Kristenson, K.J., 1959. Temperature and heat balance of soil. Oikos 10, 103–120.

Lameed, G.A., 2008. Grass-Cutter Farming in the Tropics: The Possibility and Prospects for Would-Be Farmers. Published by Peerless Prints, ISBN: 92826-47579: 96.

Lane-Petter, W., 1972. The laboratory mouse. In: UFAW (ed.), The UFAW Handbook on the Care and Management of Laboratory Animals. Williams and Wilkins, Baltimore, pp. 187–203.

Lindsey, P.A., Roviet, P.A., Romanach, S.S., 2007. Economic and conservation significance of the trophy hunting industry in Sub-Saharan Africa. Biol. Conserv. 134 (4), 455–469.

Mariki, S.B., Hassan, S.N., Maganga, S.L.S., Modest, R.B., Salehe, F.S., 2011. Wildlife-based domestic tourism in Tanzania. Ethiop. J. Environ. Stud. Manag. 4 (4), 62–73.

Martin, 2008. Review of Safari Hunting in Botswana: Financial and Economic Assessment: A Consultancy Carried Out for the Botswana Wildlife Management Association. Maun, Botswana, p. 48.

Milligain, K.R.N., Ajayi, S.S., 1978. The management of West African National Parks present status, problems and status. In: Paper presented, 4th East African Wildlife Symposium, Dec. 11–17, 1978. 31 pp.

Milligan, K.R.N., Ajayi, S.S., Hall, J.B., 1978. Density and biomass of the large herbivore community in Kainji Lake National Park. Afr. J. Ecol. 20 (1), 1–12.

Oguntala, A.B., Soladoye, M.O., 1996. Okonu Forest Reserve; Its Ecology and Management. Forestry Research Institute of Nigeria, Federal Ministry of Agriculture and Natural Resources, Ibadan.

Olawole, O.O., Ajayi, S.S., 1975. Highlights of the Problems of Bush-Meat Production and Marketing in Nigeria. vol. III Federal Department of Forestry, Nigeria. 15 pp.

Oyenuga, V.A.O., 1968. Nigeria's Food and Feeding Stuff. Ibadan University Press.

Persson, R., 1974. World Forest Resources. Skogshogskolan, Royal College of Forestry, Stockholm.

Pogosianz, H.E., 1972. The Steppe–Lemming. UFAW 292–295, Churchill & Livingstone, Edinburgh & London.

Rosevear, 1969. The Rodents of West Africa. Trustees British Museum of (Natural History), London. 604 pp.

Rowe, F.P., Taylor, E.J., Chadley, A.H.J., 1964. The effect of crowding on the production of the house mouse (Mus Musculus L.) living in Corcricks. J. Anim. Ecol. 33, 477–483.

Sanderson, I.T., 1940. The mammals of north Cameroons forest area being the results of the peray sladen expedition to the manfe division of the British Cameroons. Trans. Zool. Soc. Lond. 24, 623–725.

Simmons, R.T., Krenter, U.P., 1989. Ivory Ban; No Elephants: Policy Review. Heritage Foundation, Washington, DC, p. 2002.

Talbot, L.M., Payne, W.J.A., Ledger, H.P., Verdcourt, L.D., Talbot, M.T.T., 1965. The Meat Production Potential of Wild Animals in Africa. Comm. Agric. Bureanx, Farnhan, Royal Bucks, England.

Tewe, O.O., Ajayi, S.S., 1977. Nutritive value of the African giant rat (*Cricetomys gambianus waterhouse*). In: Paper Presented at the 8th Annual Conference, Forestry association of Nigeria, May.

Tewe, U.O., Ajayi, S.S., 1979. Replacement and nutrient value utilization by the African giant rat (Cricetomys Gambianus Waterhouse) on household waste of some local foodstuffs. J. Inst. Anim. Tech. 31 (1), 15–23.

UFAW, 1971. The rational use of living systems in bio-medical research. In: UFAW Symposium, 7th and 8th October.

UNESCO, 1972. General conference of the United Nations Education, Scientific and Cultural Organization, held in Paris 17 October to 21 November 1972.

UNESCO, 1978. UNEP and FAO Tropical Forest Ecosystems. UNESCO, Paris.

UNEP, 1995. UNEP First "Global Biodiversity Assessment" Report. United Nations Environment Programme, November 14, 1995.

ULG, 2001. ULG Economic analysis of commercial consumptive use of wildlife in Botswana. Wildlife Management Association.

WTO, 1999. Workshop on sustainable tourism indicators for Eastern and Central Europe. Keszthely (Lake Balaton) Hungary. World Tourism Organization, Madrid.

Index

Note: Page numbers followed by *f* indicate figures and *t* indicate tables.

O

P

FSC
www.fsc.org
MIX
Papier | Fördert
gute Waldnutzung
FSC® C083411

Kolibri 360 GmbH
Ferdinand-Jühlke-Straße 7
99095 Erfurt, Deutschland
produktsicherheit@kolibri360.de